# PARTICULATE AND ORGANIC MATTER FOULING OF SWRO SYSTEMS: CHARACTERIZATION, MODELLING AND APPLICATIONS

T0303875

# Particulate and organic matter fouling of SWRO systems: Characterization, modelling and applications

DISSERTATION

Submitted in fulfilment of the requirements of
the Board for Doctorates of Delft University of Technology
and of
the Academic Board of the UNESCO-IHE Institute for Water Education
for the Degree of DOCTOR
to be defended in public,
on Monday, October 3, 2011 at 10:00 o'clock
in Delft, The Netherlands

by

**Sergio Genaro SALINAS RODRÍGUEZ**

Master of Science in Water supply engineering

born in Oruro, Bolivia

This dissertation has been approved by the supervisors:

Prof. dr. G.L. Amy
Prof. dr. M.D. Kennedy


Members of the Awarding Committee:

Rector Magnificus, Delft University of Technology, chairman
Prof. dr. A. Szöllösi-Nagy, UNESCO-IHE, vice-chairman
Prof. dr. G.L. Amy, UNESCO-IHE/Delft University of Technology, supervisor
Prof. dr. M.D. Kennedy, UNESCO-IHE/Delft Univ. of Technology, supervisor
Prof. dr. ir. J.C. Schippers, UNESCO-IHE/Wageningen University
Prof. dr. B. Mariñas, University of Illinois
Prof. dr. E. Drioli, University of Calabria
Prof. dr. ir. J.C. van Dijk, Delft University of Technology
Prof. dr. ir. L.C. Rietveld, Delft University of Technology, reserve

CRC Press/Balkema is an imprint of the Taylor & Francis Group, an informa business

Published by:
CRC Press/Balkema
PO Box 447, 2300 AK Leiden, the Netherlands
e-mail: Pub.NL@taylorandfrancis.com
www.crcpress.com - www.taylorandfrancis.co.uk - www.ba.balkema.nl

ISBN 978-0-415-62092-5 (Taylor & Francis Group)

# Foreword

The research presented in this thesis has been carried out at UNESCO-IHE, as a part of the EU-MEDINA fp6 project "Membrane based desalination – An integrated approach" in which 13 partners – universities, research institutes and industry – were involved from December 2006 until 2010.

I wish to express my sincere gratitude to my supervisors, Gary Amy, Maria Kennedy and Jan Schippers, for granting me their strong support, guidance and encouragement to perform research – in the field of membrane technology and desalination – in the most motivating way.

A doctoral degree is not achieved on one's own. Colleagues are there to provide technical support, discussions, discerning comments and warmth. Bearers of these have been (in alphabetical order) Andrew Maeng, Arpad Gonzales, Berry Gersonius, Don van Galen, Eduard Gasia, Frans Oostrum, Fred Kruis, Frederik Spenkelink, Jan Herman Koster, Javier Patarroyo, Jolanda Boots, Loreen Villacorte, Lyzette Robbemont, Marcelo Gutiérrez, Mariette Halkema, Miguel Gutiérrez, Rinnert Schurer, Saeed Baghoth, Sophie Rapenne, Stefan Huber, Steven Mookhoek, Tanny van der Klis, Tarek Waly, Theo van der Kaaij, Ton Keijzer, and Victor Yangali. Four master students contributed to this research: Badar Al-Rabaani, Mamoun Althuluth, Didik Wahyudi and Yuli Ekowati.

Last but not least, I thank to my parents and brothers for their encouragement, love and support during all my stay abroad. Mariska, this would have not been the same without you. Thank you for your love, help and patience. I am looking forward to the future years as a family!

Sergio Salinas R.

Delft, June 2011

A mi pá

*"Soy feliz, qué carajo"*

# Contents

# Chapter 1

## 1   Introduction

## 1.1   Introduction

Desalination is increasingly being touted as a solution to the world water crisis in the 21st century. Considering that almost one quarter of the world's population lives less than 25 km from the coast, seawater could become one of the main sources of freshwater in the near future (Drioli and Macedonio, 2010, ETAP, 2006). Reverse osmosis (RO), a technique pioneered in the second half of the 20th century, has become a basis of water production in many parts of the Middle East, North Africa, Australia and Europe. The desalination market in these parts of the world is currently in the growth stage of its life cycle. This growth is fuelled by a need for potable water not only for human consumption, but also for irrigation, industrial and tourism purposes.

According to a report from Pike Research (2010), factors like water scarcity, population and economic growth, pollution, and urbanization will contribute to strong growth in the desalination technology market over the next several years, and is forecasted that global desalination investment will double from $8.3 billion in 2010 to $16.6 billion per year by 2016, representing cumulative spending of $87.8 billion during that period (Clean edge, 2011). The market has been growing at a fluctuating rate over recent years due to its dependence on large plants (Figure 1.1).

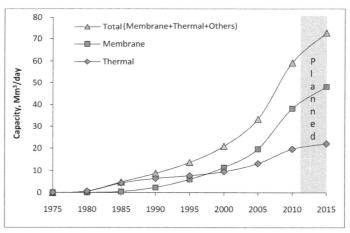

**Figure 1.1. Production capacity in the world over time for Membrane-based and Thermal desalination (DesalData, 2011)**

In Figure 1.2 it is shown for selected countries the increase in total desalination capacity from 2007 until 2011. The growth in membrane applications for water surface treatment, and brackish and seawater desalination has been significant since 1990. This growth in membrane applications has resulted in a decreasing cost of the desalination facilities, with the consequence that the unit cost of the product water from membrane plants has been also lowered, even for the very energy-intensive thermal plants in the Gulf region - which purify seawater by boiling and condensing -

can produce fresh water at less than US$1 per cubic metre (Schiermeier, 2008). For instance, the desalination plant at Ashkelon (Israel), produces more than 300,000 $m^3$ of freshwater per day at costs of around US$ 50 cents/$m^3$. Furthermore, in Table 1.1 the costs indications to produce one cubic metre of water is presented for various technologies.

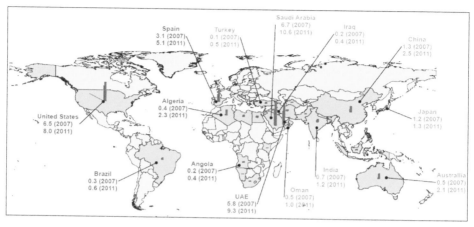

**Figure 1.2. Desalination capacity in the world for 2007 and 2011 in Mm³/d (Based on (DesalData, 2011))**

**Table 1.1. Cost indications**

| Technology | €/m³ |
|---|---|
| Seawater reverse osmosis | 0.50 – 1.00 |
| Brackish water reverse osmosis | 0.25 – 0.50 |
| Electrodialysis | 0.25 – 0.50 |
| Nanofiltration | 0.15 – 0.25 |
| Ultra/microfiltration | 0.05 – 0.10 |

(Schippers, 2010)

In Table 1.2, the bigger seawater membrane-based desalination plants in the world are listed which are currently operating, under construction or that are planned to operate in the coming five to eight years. The listed plants currently under construction will provide an additional 8,85 $Mm^3$/d to the current desalination capacity in the world.

Desalination in the broad sense is a process through which water of low salinity is produced to an extent that it becomes potable. Among the known desalination processes, multistage flash (MSF) distillation and reverse osmosis (RO) membrane filtration are most popular and widely used techniques. Major chemical constituents (35,000-50,000 mg/L) of seawater are of inorganic origin and the minor (1-4 mg/L) are of organic origin. Though organics are negligible in concentration as compared to inorganic constituents, they pose more acute problems in reverse osmosis desalination process. It is well known that fouling in RO membranes causes serious problems including (i) a gradual decline of membrane flux thereby decrease in permeate production, (ii) an increase in pressure thereby increasing requirement of high pressure pump rating and (iii) degradation of membrane itself. All these factors reflect on the cost of water production. Hence, nowadays attempts are being made to deplete the concentration of organic and some of the inorganic constituents from the feed to RO to overcome these problems by various pre-treatment methods. Other than conventional methods such as coagulation (Croué et al., 1993), filtration and separately passing through activated carbon or clays for decreasing the organic load from the feed of RO, some of the more recent techniques (Alborzfar et al., 1998, Kati et al., 1998) are alternatives such as nanofiltration (NF) and ultrafiltration (UF). Moreover, these organic contaminants have been found to be the precursors for the formation of organic derivatives, some of which are carcinogenic.

**Table 1.2. Seawater reverse osmosis plants with bigger capacity in the world (DesalData, 2011)**

| Project Name | Country | Capacity ($m^3/d$) | Use | Contract date | Online date | Status |
|---|---|---|---|---|---|---|
| Jordan Red Sea Project Phase 1 | Jordan | 575385 | DW | | 2018 | Planned |
| Soreq | Israel | 510000 | DW | 2009 | 2013 | Constr. |
| Tripoli East | Libya | 500000 | DW | | | Planned |
| Misurata | Libya | 500000 | - | | | Planned |
| Mactaa | Algeria | 500000 | DW | 2008 | 2011 | Constr. |
| Hamriyah IV | UAE | 455000 | DW | | | Planned |
| Wonthaggi | Australia | 444000 | DW | 2009 | 2011 | Constr. |
| Soreq 2 | Israel | 411000 | DW | 2010 | 2012 | Planned |
| Benghazi | Libya | 400000 | - | | | Planned |
| Ashkelon | Israel | 326144 | DW | 2002 | 2005 | *Online* |
| Ashdod | Israel | 320000 | DW | 2009 | 2013 | Constr. |
| Tuas II | Singapore | 318500 | DW | 2011 | 2013 | Constr. |
| Ras Azzour (RO) | KSA | 306686 | DW | 2010 | 2014 | Constr. |
| Gulf of Mexico to supply San Antonio, TX | USA | 302800 | DW | | | Planned |
| Kerman | Iran | 300000 | DW | | | Planned |
| Jfara | Libya | 300000 | - | | | Planned |
| Lima North | Peru | 300000 | DW | | | Planned |
| Istanbul | Turkey | 300000 | DW | | | Planned |
| Port Stanvac | Australia | 300000 | DW | 2009 | 2012 | Constr. |
| Ashdod (Paz Oil Company) | Israel | 274000 | - | | 2013 | Planned |
| Hadera | Israel | 272765 | DW | 2006 | 2010 | *Online* |
| Chtouka | Morocco | 250000 | Irr. | | | Planned |
| Sydney (Kurnell) | Australia | 250000 | DW | 2007 | 2010 | *Online* |
| Torrevieja, Alicante/Murcia | Spain | 240000 | Irr. | 2006 | 2011 | *Online* |
| Jeddah Phase 3 | KSA | 240000 | - | 2008 | 2012 | Constr. |
| Korangi | Pakistan | 227300 | DW | 2008 | | Planned |
| Ad Dur | Bahrain | 218208 | DW | 2008 | 2015 | Constr. |
| Rabigh IWSPP | KSA | 218000 | Ind. | 2005 | 2008 | *Online* |
| Shuqaiq 2 | KSA | 213000 | DW | 2006 | 2010 | *Online* |
| Laoting | China | 200000 | Ind. | 2011 | | Planned |
| Chennai 3 | India | 200000 | DW | | | Planned |
| Mundra SEZ expansion | India | 200000 | Pow. | | | Planned |
| El Prat de Llobregat | Spain | 200000 | DW | 2006 | 2009 | *Online* |

DW=Drinking water, Irr.=Irrigation, Ind.=Industry, Pow.=Power station

Feed water for membrane based desalination plants needs extensive pre-treatment in order to prevent membrane fouling of spiral wound reverse osmosis elements. Pre-treatment is one of the factors determining the success or failure of a desalination installation, and influences the overall performance of the plant. Conventional pre-treatment is based on mechanical treatment (media filters, cartridge filters) supported by an extensive chemical treatment, including biofouling control (chlorination, dechlorination), removal of suspended material (flocculant dosing), and scaling prevention (dosing of acids or generic antiscalant additives). Specific additives have to be used for the preservation of the RO membranes during storage and transport. Seasonal variations in seawater quality further cause difficulties in process control. This may increase the frequency of chemical cleaning to prevent efficiency loss in the process. Cleaning may involve alkaline solutions (pH 11-12) for removal of silt deposits and biofilms, or acidic solutions (pH 1-2) to dissolve metal oxides or scales, as well as wide range of additional chemicals such as detergents, oxidants, complexing agents and biocides. As a result, the pre-treatment and cleaning may account for a significant part of the total cost (Van der Bruggen and Vandecasteele, 2002). Furthermore, as the pre-treatment and cleaning chemicals are typically discharged along with the concentrate into surface waters or into the sea, they are a central aspect in the discussion of environmental impacts (Lattemann, 2010, Lattemann and Hoepner, 2003).

Pressure driven membrane processes (microfiltration, ultrafiltration, nanofiltration) are an alternative to conventional pre-treatment in designing pre-treatment systems. Microfiltration (MF) is an obvious technique for the removal of suspended solids and for lowering the silt density index (SDI). Energy consumption in MF is relatively low, whereas the cost for a corresponding conventional pre-treatment is more than double (Van der Bruggen and Vandecasteele, 2002, Ebrahim et al., 2001). MF generally provides an RO feedwater of good quality, with a lower SDI in comparison to the untreated seawater, although there is a large influence of the feedwater quality. Further improvement of the RO feedwater can be obtained by replacing and/or adding MF by ultrafiltration (UF). In UF, not only suspended solids and large bacteria are retained, but also (dissolved) macromolecules, colloids and smaller bacteria. A further removal of organics can be achieved by combining MF or UF with a coagulation step before. Because of higher applied pressure, UF cost is higher than that for MF, but competitive with conventional pre-treatment. On the other hand, the UF permeate (the RO feed) is significantly improved.

According to Potts et al. (Potts et al., 1981) and references cited therein, particulate matter in natural waters and waste waters can be classified as settleable solids (>100 µm), supra-colloidal solids (1 µm to 100 µm), colloidal solids (0.001 µm to 1 µm) and dissolved solids (<0.001 µm). The above cut-offs are more or less arbitrary and different values are set by different authors. The chemical composition of particulate solids is of a wide variety and a major distinction is between inorganic and organic matter. The most common inorganic particles are aluminium silicate clays, ranging in size between 0.3 and 1 μm, and colloids of iron, aluminium and silica. Organic particles include proteins, carbohydrates, fats, oils and greases, and various surfactants. Polyphenolic aromatic complexes such as humic acids, lignin and tannin are decay products of woody tissues of plants and often occur as very small colloids. Polysaccharides that constitute cell walls of microorganisms and plants are also prominent. Some of the types of colloids that exist in natural waters, especially in the sea, are listed in Table 1.3 (Ning, 1999).

**Table 1.3. Colloidal matter in natural waters**

Microorganisms
Biological debris (plant and animal)
Polysaccharides (gums, slime, plankton, fibrils)
Lipoproteins (secretions)
Clay (hydrous aluminium and iron silicates)
Silt
Oils
Kerogen (aged polysaccharides, marine snow)
Humic acids, lignins, tannins
Iron and manganese oxides
Calcium carbonate
Sulphur and sulphides

Fouling represents the major constraint to more cost-effective, and therefore expanded, application of membrane technology in drinking water, particularly for reverse osmosis systems. Fouling can occur in several forms:

- *Particulate fouling:* particles and colloids not retained by pre-treatment impart resistance and reduce flux (in constant pressure filtration) or demand higher feed pressure (in constant flux filtration) as particle deposits accumulate onto the membrane surface.

- *Organic fouling:* is associated with bulk organic matter (OM) present in the feed water passing through the pre-treatment processes and that may be adsorbed onto the membrane surface as a gel-layer. In addition to macromolecules, organic foulants can include organic colloids (Amy, 2008). Moreover, the biodegradable organic matter (BOM) retained on the membrane surface can be utilized by micro-organisms as nutrients and may contribute to biological growth.

- *Biofouling:* microorganisms tend to adhere to surfaces (e.g., membrane surface) and to form a gel layer called biofilm. On the raw water side, the biofilm causes an increase of fluid friction resistance which increases the differential feed/concentrate pressure. Also, overall hydraulic resistance of the membrane can increase due to the biofilm. If these effects exceed a certain threshold of interference, they are addressed as biofouling (Flemming et al., 1997).

- *Scaling:* salt precipitation occurs on the surface of the membrane due to localized supersaturation conditions.

It is clear therefore that the distinction between particulate/ colloidal, organic, biological fouling and scaling of membranes is not sharp but there exists a certain degree of overlap. However, this distinction is a useful one since methods of fouling prediction, prevention through pre-treatment and operating measures, as well as fouling deposit removal through cleaning, are categorized in the same manner (Yiantsios et al., 2005).

Reliable methods to predict the fouling potential of reverse osmosis (RO) feed water are important in preventing and diagnosing fouling at the design stage, and for monitoring pre-treatment performance during plant operation (Boerlage, 2007). Particles and colloids (both inorganic and organic) are one of the possible origins of RO fouling. However, as mentioned by many researchers, fouling is complex and it may be due to several contributing factors.

Traditionally, the *NOM fouling* potential of a feed water has been assessed in terms of dissolved organic carbon (DOC), UV absorbance, and colour; however, NOM fouling rates do not appear to correlate with these traditional water quality parameters. A problem is that DOC only indicates the amount but not the character of the NOM. More recently, specific UV absorbance (SUVA) has been used to indicate the aromatic character of NOM but SUVA is a direct measure of humic substances which are less problematic as foulants compared to non-humic materials. A quality parameter that can describe the level of organic fouling expected for a particular membrane feed water (surface, brackish and seawater) is urgently required. There has been no surrogate test developed for indirect assessment of NOM fouling potential.

In membrane based desalination systems, it is common practice to judge or assess the quality of a feed/pre-treated water via traditional fouling indices such as silt density index (SDI) and modified fouling index (MFI ($MFI_{0.45}$). For both tests, membranes with pores of 0.45 μm are used and flux decline is measured at constant pressure. However, alternatively, MFI tests can also be done with membranes of different pore sizes and at constant flux. Many studies have reported that the main difficulty with SDI, is the lack of reproducible results when performing the tests with various membrane materials and even within the same batch of manufactured filters. SDI shows several deficiencies, e.g., no linear relation with concentration of suspended and colloidal matter; no correction for temperature; and it is not based on any filtration mechanism. $MFI_{0.45}$ (based on 0.45 μm filtration) is a superior alternative since it: shows a linear relation with particle concentration; is corrected for temperature; and is based on the cake filtration mechanism. However, one of the main deficiencies in existing fouling indices (SDI and $MFI_{0.45}$) is that they operate at constant pressure producing high initial flux values. Boerlage et al. (1997, 2001) further developed the concept of MFI by using ultrafiltration (UF) membranes and initially proposed the use of constant flux filtration instead of constant pressure filtration.

Both SDI and $MFI_{0.45}$ have no value in predicting the rate of fouling due to particle deposition on RO/NF membrane surfaces. Both might have predictive value in clogging, e.g., non-woven fabric and fibre bundles in DuPont's and Toyobo's permeators and spacers of spiral wound elements. This is the main motivation for development of the MFI-UF – measured with membranes of different pore sizes.

## 1.2   Aims and scope

Fouling in any of its forms is a limiting factor in the control and operation of reverse osmosis plants. It is acknowledged in practice that the control of organic matter and particulate fouling is fundamental in decreasing costs related with membrane filtration applications independently of the applications.

The control of organic matter and particulate fouling in membrane filtration systems can be improved by a clearer understanding of the processes involved in these phenomena and more accurate methods to predict and prevent these phenomena.

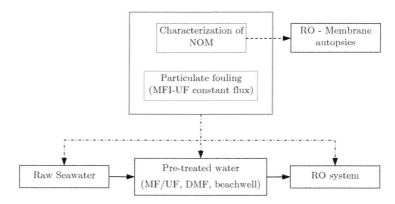

**Figure 1.3. Approach of the project**

Figure 1.3 depicts the research methodology integrated into a desalination process scheme which has been the purpose of this study.

This research focus on the further development and applications of water quality assessment tools to directly and/or indirectly predict the fouling potential of feed waters applied to seawater and estuarine water reverse osmosis. Characterisation of natural organic matter in seawater by two analytical techniques (i.e., liquid chromatography with organic carbon detection and mapping through 3D fluorescence) and the further development of the modified fouling index with ultrafiltration membranes at constant flux filtration are the main focus of this research.

The final goal is to achieve a better knowledge of organic matter fouling and particulate matter fouling, that should enable engineers, plant operators and scientists not only to design better plants, but also to develop more effective tools for operation and monitoring of fouling.

## 1.2.1  RESEARCH OBJECTIVES

The main objectives of the research are the following:

- To characterize the natural organic matter in seawater and estuarine water by various analytical techniques and link these measurements with fouling in membrane systems with different pre-treatment.
- To further develop the modified fouling index with ultrafiltration membranes at constant flux filtration as an accurate tool to assess pre-treatment and a tool to estimate the rate of particulate/ colloidal fouling in seawater reverse osmosis systems.

## 1.3  Outline of the study

This thesis contains nine chapters. The first chapter corresponds to the introduction of the study. The last chapter summarizes the major conclusions of the study.

There are two chapters (2 and 3) dedicated to organic matter characterization in seawater. Chapter 2 deals with the testing protocols and applications for mapping of organic matter components through liquid chromatography and fluorescence spectroscopy under high ionic strength conditions including parallel factor analysis and principal components analysis for seawater and estuarine water samples.

Chapter 3 makes use of the laboratory techniques described in chapter 2 to identify organic foulants in seawater, estuarine and bay sources for reverse osmosis plants. Several locations in Europe were studied.

Particulate/colloidal fouling potential is studied in the other five chapters (4, 5, 6, 7 and 8). Chapter 4 is the introduction to particulate/ colloidal fouling indices and presents a review of the current status of fouling indices used in seawater RO systems. Indices such as: Silt Density Index (SDI), Modified Fouling Index (MFI), MFI-UF constant pressure, MFI-UF constant flux and cross flow sampler (CFS) coupled with MFI-UF are discussed.

Chapter 5 presents the set-up and method development and applications related to the modified fouling index with ultrafiltration membranes at constant flux filtration. The chapter characterizes the proposed membranes, describes the testing procedure for MFI-UF constant flux measurements and defines the limit of detection of the test. Applications related to comparison of various raw waters, particle size distribution, plant profiling, pre-treatment assessment and RO particulate fouling prediction are presented.

Chapter 6 studies the effect of flux rate on cake compression and on arrangement of particles in membrane filtration and on fouling indices.

Chapter 7 studies the particle deposition/accumulation in seawater reverse osmosis systems by measuring the particle deposition factor based on the MFI-UF constant flux measurements. A correction factor is proposed to consider effect of ionic strength on MFI values of RO concentrate.

Chapter 8 presents applications of the MFI-UF constant flux in pre-treatment assessment and in particulate fouling prediction.

Most of the results presented in this dissertation have been published in journals, presented in conferences, or have been submitted for publication in specialized journals.

The intention of each chapter is that it should present all necessary information by itself; in this sense, the reader may find duplicated information, particularly regarding to the material and methods followed in the study and in some cases results are repeated.

## 1.4   References

ADHAM, S. & FANE, A. 2008. Cross Flow Sampler Fouling Index. California, USA: National Water Research Institute.

AL-HADIDI, A. M. M. 2011. *Limitations, Improvements, Alternatives for the Silt Density Index*, Enschede, Gildeprint Drukkerijen.

ALBORZFAR, M., JONSSON, G. & GRON, C. 1998. Removal of natural organic matter from two types of humic ground waters by nanofiltration. *Water Research*, 32, 2983-2994.

AMY, G. 2008. Fundamental understanding of organic matter fouling of membranes. *Desalination*, 231, 44-51.

BOERLAGE, S. F. E. 2001. *Scaling and Particulate Fouling in Membrane Filtration Systems*, Lisse, Swets&Zeitlinger Publishers.

BOERLAGE, S. F. E. 2007. Understanding the SDI and Modified Fouling Indices (MFI0.45 and MFI-UF). *IDA World Congress On Desalination and Water Reuse 2007 - Desalination: Quenching a Thirst* Maspalomas, Gran Canaria - Spain.

BOERLAGE, S. F. E., KENNEDY, M. D., BONNE, P. A. C., GALJAARD, G. & SCHIPPERS, J. C. 1997. Prediction of flux decline in membrane systems due to particulate fouling. *Desalination,* 113, 231-233.

CLEAN EDGE. 2011. Desalination Plants to Attract $87.8 Billion in Investment by 2016. *Clean edge news* [Online]. [Accessed 21.08.11].

CROUÉ, J. P., LEFEBRE, F., MARTIN, B. & LEGUBE, B. 1993. Removal of dissolved hydrophobic and hydrophilic organic substances during coagulation/flocculation of surface waters. *Water Science Technology,* 27, 143-152.

DESALDATA 2011. IDA Desalination Plants Inventory. *In:* GLOBAL WATER INTELLIGENCE & WATER DESALINATION REPORT (eds.).

DRIOLI, E. & MACEDONIO, F. 2010. New Trends in Membrane Technology for Water Treatment and Desalination.

EBRAHIM, S., ABDEL-JAWAD, M., BOU-HAMAD, S. & SAFAR, M. 2001. Fifteen years of R&D program in seawater desalination at KISR Part I. Pretreatment technologies for RO systems. *Desalination,* 135, 141-153.

ETAP 2006. Water Desalination Market Acceleration. *Environmental Technologies Action Plan,* http://ec.eruopa.eu.environment/etap.

KATI, R., VAISANEN, P., METASA, M. S., KUTOVAARA, M. & NYSTROM, M. 1998. Characterization and removal of humic substances in Ultra and nanofiltration. 118, 273-283.

LATTEMANN, S. 2010. *Development of an environmental impact assessment and decision support system for seawater desalination plants,* Delft, CRC Press/Balkema.

LATTEMANN, S. & HOEPNER, T. 2003. *Seawater Desalination - Impacts of Brine and Chemical Discharges on the Marine Environment,* L'Aquila, Italy Desalination Publications.

NING, R. Y. 1999. Reverse osmosis process chemistry relevant to the Gulf. *Desalination,* 123, 157-164.

PIKE RESEARCH. 2010. Desalination Technology Markets. [Accessed 21.08.11].

POTTS, D. E., AHLERT, R. C. & WANG, S. S. 1981. A critical review of fouling of reverse osmosis membranes. *Desalination,* 36, 235-264.

SCHIERMEIER, Q. 2008. Purification with a pinch of salt. *Nature,* 452, 260-261.

SCHIPPERS, J. C. 2010. Introduction to membrane technology. *Membrane Technology in Drinking & Industrial Water Treatment. Principles, Design & Applications.* Delft, The Netherlands: Unesco-IHE.

SCHIPPERS, J. C. & VERDOUW, J. 1980. The modified fouling index, a method of determining the fouling characteristics of water. *Desalination* 32, 137-148.

SIM, L. N., YE, Y., CHEN, V. & FANE, A. G. 2010. Crossflow Sampler Modified Fouling Index Ultrafiltration (CFS-MFIUF) – An alternative fouling index. *Journal of Membrane Science,* 360, 174-184.

SIM, L. N., YE, Y., CHEN, V. & FANE, A. G. 2011a. Comparison of MFI-UF constant pressure, MFI-UF constant flux and Crossflow Sampler-Modified Fouling Index Ultrafiltration (CFS-MFIUF). *Water Research,* 45, 1639-1650.

SIM, L. N., YE, Y., CHEN, V. & FANE, A. G. 2011b. Investigations of the coupled effect of cake-enhanced osmotic pressure and colloidal fouling in RO using crossflow sampler-modified fouling index ultrafiltration. *Desalination,* 273, 184-196.

SIOUTOPOULOS, D. C., YIANTSIOS, S. G. & KARABELAS, A. J. 2010. Relation between fouling characteristics of RO and UF membranes in experiments with colloidal organic and inorganic species. *Journal of Membrane Science,* 350, 62-82.

VAN DER BRUGGEN, B. & VANDECASTEELE, C. 2002. Distillation vs. membrane filtration: overview of process evolutions in seawater desalination. *Desalination,* 143, 207-218.

YIANTSIOS, S. G., SIOUTOPOULOS, D. & KARABELAS, A. J. 2005. Colloidal fouling of RO membranes: an overview of key issues and efforts to develop improved prediction techniques. *Desalination,* 183, 257-272.

YU, Y., LEE, S., HONG, K. & HONG, S. 2010. Evaluation of membrane fouling potential by multiple membrane array system (MMAS): Measurements and applications. *Journal of Membrane Science,* 362, 279-288.

# Chapter 2

# 2 Mapping of organic matter components through liquid chromatography and fluorescence spectroscopy under high ionic strength conditions

Chapter 2 is based on:

SALINAS RODRÍGUEZ, S. G., GONZALES T, A., KENNEDY, M. & AMY, G. (2008). Fluorescence of selected organic matter compounds: looking at the effect of concentration, ionic strength and pH. *In:* BIRMINGHAM, U. O., ed. AGU Chapman. *Conference on Organic Matter Fluorescence*, Edgbaston, Birmingham, UK.

SALINAS RODRÍGUEZ, S. G., KENNEDY, M. D., SCHIPPERS, J. C., & AMY, G. (2011). Mapping of organic matter components through liquid chromatography and fluorescence spectroscopy under high ionic strength conditions. *Desalination*, submitted.

LOZIER, J. C., BANKSTON, A., BEATY, J., GARCIA-ALEMAN, J., SCHARF, R., AMY, G. & SALINAS RODRÍGUEZ, S. G. (2009). Use of advanced NOM characterization methods to trace the fate of organic contaminants from a membrane backwash recycle scheme. *In:* AWWA, ed. *Membrane Technology Conference*, Memphis, Tennessee United States. AWWA and AMTA.

## 2.1   Introduction

Fouling represents the major constraint to more cost-effective, and therefore expanded, application of membrane technology in drinking water treatment, particularly for reverse osmosis systems. Fouling can occur in several forms: particulate, organic, biological, and inorganic fouling, and scaling.

Traditionally, the fouling potential of a feed water due to natural organic matter (NOM) has been assessed in terms of dissolved organic carbon (DOC), UV absorbance, and color; however, NOM fouling rates do not appear to correlate with these traditional water quality parameters. A problem is that DOC only indicates the *amount* but not the *character* of the NOM. More recently, specific UV absorbance (SUVA) has been used to indicate the aromatic character of NOM but SUVA is a direct measure of humic substances which are less problematic as foulants compared to non-humic materials. A quality parameter that can describe the level of organic fouling expected for a particular membrane feed water (surface, brackish and seawater) is urgently required. There has been no surrogate test developed for indirect assessment of NOM fouling potential. Nevertheless, in the last couple of years transparent exopolymer particles have been linked to operational problems in integrated membrane systems (Berman and Passow, 2007, Villacorte et al., 2009a, Villacorte et al., 2009b).

Liquid chromatography has been extensively applied to monitor changes in organic matter through water treatment processes. Fluorescence excitation–emission measurements are used in many different fields as described by Bro and Vidal (2010) such as skin analysis, fermentation monitoring, environmental analysis, food and clinical analysis. Common to many of these fields is that the spectroscopic measurements are performed directly on complex mixtures rather than on simple purified samples. This has posed severe problems in the subsequent analysis of the data, because the contributions of the individual fluorophores usually overlap. Multivariate methods have made it possible to develop calibration models for specific properties but not to extract the full information available in the data.

In this chapter, liquid chromatography with organic carbon detection (LC-OCD) and fluorescence excitation-emission matrix (F-EEM) techniques are evaluated under high ionic strength conditions and their testing protocols adapted, if needed, to seawater and brackish water. Standard organic matter compounds representative of proteins, amino acids, humic substances and reference materials were used to tests the test protocols and at the same time these compounds were mapped by these two techniques. A dataset of F-EEMs samples was modeled by parallel factor analysis (PARAFAC) and the LC-OCD results from the same samples was subjected to principal component analysis (PAC). A dataset was created with samples from different locations along the Mediterranean Sea and North Sea.

## 2.2   Materials and methods

Two methods were studied, namely liquid chromatography coupled with organic carbon detection and fluorescence spectroscopy. With LC-OCD organic matter fractions are classified by size. With fluorescence, two distinct classes of fluorophores are generally delineated, the humic like fluorophores and the protein-like fluorophores. For example, Coble referred to two humic-like fluorophores in addition to a specific marine humic-like fluorophore and one or two protein-like fluorophores (tyrosine and tryptophan-like), depending on the origin of the water samples (Leenheer and Croué, 2003).

### 2.2.1   LIQUID CHROMATOGRAPHY - ORGANIC CARBON DETECTION

Liquid chromatography - organic carbon detection (LC-OCD) (also called size exclusion chromatography – organic carbon detection, SEC-DOC) can be used to effectively monitor NOM components with a lower SUVA. LC-OCD has been successfully applied to monitoring changes in NOM associated with fresh water treatment (Her et al., 2002), and has also been used to identify problematic NOM components in membrane fouling (Her et al., 2004). LC-OCD separates NOM according size/molecular weight (MW) classes ranging from higher to lower MW: biopolymers (BP), humic substances (HS), building blocks (BB), low MW acids (LMA) and neutrals (N). The magnitude of the BP peak has been linked with fouling potential in UF membranes (Amy and Her, 2004).

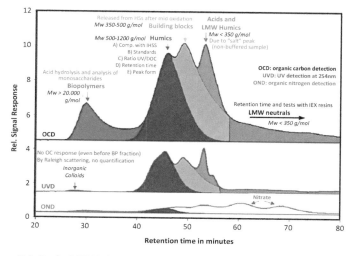

Figure 2.1. Typical NOM chromatogram of a fresh water sample [Huber (2007)]

A typical chromatogram of NOM contained in surface water is illustrated in Figure 2.1. The first fraction identified after a retention time of approximately 25 - 45 minutes (first peak – largest molecular size) is the biopolymer peak with significant response by organic carbon detection (OCD) only. The organic colloids and proteins present in this fraction provide response in OCD and UV detection. The second and third fraction responses in OCD and UVD are attributed to humic substances and building blocks, respectively. Building blocks are a weathering product of humic substances. The fourth and fifth responses to OCD and UV detection is attributed to low MW acids and neutrals, respectively.

In this study, the laboratory facilities of *Hetwaterlaboratorium* (Haarlem, The Netherlands), were used for running the LC-OCD analysis of water samples. In all cases, the chromatograms were resolved with the help of Fiffikus software (DOC-Labor Dr. Huber).

## 2.2.2   FLUORESCENCE EXCITATION-EMISSION MATRIX

Fluorescence spectroscopy has low detection levels compared to other spectroscopic techniques (Bro and Vidal, 2010, Stedmon et al., 2003). This makes it a valuable technique for measuring trace concentrations in products. In low concentrations there is a linear relationship between the measured signal and the concentration of the fluorophore (Skoog and Leary, 1992). Deviations from this linear relationship may be caused by high concentrations of the fluorophore itself, causing inner filter effect and/or quenching.

A fluorescence excitation-emission matrix (F-EEM) is developed by scanning over an excitation range of 240 to 450 nm by 10-nm increments and an emission range of 290 to 530 nm by 2 nm increments using a FluoroMax-3 spectrofluorometer (HORIBA Jobin Yvon, Inc., USA). The fluorescence samples are first adjusted to ~1 mg/L of DOC and a pH of ~2.8. The result is a three-dimension spectrum in which fluorescence intensity (normalized to Raman units, RU) is represented as a function of excitation and emission wavelengths.

**Table 2.1. Typical EEM peak values**

| Description | Wavelength, nm | |
| --- | --- | --- |
| | Ex | Em |
| Humic-Like primary peak (humic peak) | 330–350 | 420–480 |
| Humic-Like secondary peak (fulvic peak) | 250–260 | 380–480 |
| Marine humic-like peak | 310-330 | 400-420 |
| Amino acid-Like (Tyrosine) peak | 270–280 | 300–320 |
| Amino acid-Like (Tryptophan) peak | 270–280 | 320–350 |
| Protein-Like (Albumin) peak | 280 | 320 |

In order to further remove the Rayleigh scattering effects, emission measurement data made in the region of the excitation wavelength $\pm$ 20 nm were deleted, and a set of zeros were inserted in a triangular-shaped region where the emission wavelength is less than excitation wavelength (upper corner of left side of EEM).

Protein-like organic matter, hypothesized to be a principal membrane foulant (polysaccharides are potential foulants as well), exhibits a dominant peak at lower excitation/emission wavelengths while humic/fulvic substances show dominant primary and secondary peaks at higher excitation/emission wavelengths. Table 2.1 shows typical responses for various compounds.

Based on an F-EEM, a fluorescence index (FI) can be calculated by the ratio of fluorescence intensity at emission 450 and 500 nm at excitation 370 nm. A higher FI (~1.7-~2.0) reflects organic matter of an autochthonous (microbial) origin while a lower FI (~1.3-~1.4) reflects organic matter of an allochthonous (terrestrial) origin (McKnight et al., 2001).

## 2.2.3  STANDARD ORGANIC MATTER COMPOUNDS

The standard organic matter compounds that were studied are summarized in Table 2.2.

Table 2.2. Standard organic matter compounds used in the study

| Type | Name | Formula | Molecular weight | Comment |
|---|---|---|---|---|
| Humic Substance | Suwannee River Humic Acid (HA) | - | 0.5-5 kDa | Blackwater river. Produced by IHSS* |
| | Suwannee River Fulvic Acid (FA) | - | 0.5-5 kDa | Blackwater river. Produced by IHSS* |
| | Aldrich humic acid (AHA) | - | $<10$ - $>100$ kDa** | Soil based. |
| Amino acid | Tyrosine | $C_9H_{11}NO_3$ | 181.19 g/mol | Used by cells to synthesize proteins |
| | Tryptophan | $C_{11}H_{12}N_2O_2$ | 204.23 g/mol | Used in structural or enzyme proteins |
| Protein | Albumin | - | ~66.4 kDa | Bovine serum albumin |
| Deep-Sea reference material | Sargasso Sea & Florida Strait | - | - | Purchased from the University of Miami |
| Alginate | Alginic acid sodium salt | $NaC_6H_7O_6$ | 12-80 kDa | Low molecular weight |
| | Dansyl Alginate | $[C_{18}H_{19}NO_8S]_n$ | ~350 kDa | High molecular weight |
| Peptidoglycan | Peptidoglycan from staphylococcus aureus | - | - | Purchased from Aldrich |
| Quinone | 1,4-Benzoquinone | $C_6H_4O_2$ | 108.09 g/mol | Purchased from Aldrich |

*IHSS = International humic substances society
** Apparent molecular weight: 18 % $<$ 10 kDa, 20 % 10-30 kDa, 30 % 30-100 kDa, and 25 % >100 kDa (Katsoufidou et al., 2008, Sioutopoulos et al., 2010)

The Suwannee River is a blackwater river, with DOC concentrations ranging from 25-75 mg/L and pH values of less than pH 4.0. The precise properties and structure of a given humic substance sample depends on the water or soil source and the specific conditions of extraction. Nevertheless, the average properties of HA, FA and humin from different sources are remarkably similar (IHSS, 2010). Humic substances have been regarded as macromolecular, but recent studies of aqueous humic extracts from soil (Simpson et al., 2002), lignite (Piccolo et al., 2002), and water (Leenheer et al., 1989, Leenheer et al., 2001) found relatively small primary molecular structures (100–2000 Da) with macromolecular characteristics resulting from aggregates formed by hydrogen bonding, nonpolar interactions, and polyvalent cation interactions.

Most bacteria have a cell wall containing a special polymer called peptidoglycan. Over the cell membrane is a shift of peptidoglycan and other polymers including teichoic and teichuronic acids. The bacterial cell is a unique biopolymer, and contains both D- and L-amino acids (Sigma-Aldrich, 2010).

Albumin is the main protein of plasma; it binds water, cations (such as $Ca^{2+}$, $Na^+$ and $K^+$), fatty acids, hormones, bilirubin and drugs - its main function is to regulate the colloidal osmotic pressure of blood.

A quinone is a class of organic compounds that are formally derived from aromatic compounds (such as benzene or naphthalene) by exchanging an even number of –CH= groups by –C(=O)– groups, with any necessary rearrangement of double bonds, resulting in a fully conjugated cyclic dione structure. More recently, quinone moieties resulting from phenol oxidation were found to contribute significantly to the fluorescence of humic substances extracted from marine sediments (Klapper et al., 2002).

## 2.3   Fluorescence – Excitation emission matrix

Fluorescence spectroscopy is widely used to monitor environmental changes in aquatic environments such as lakes, rivers and seas. There are many studies that have reported applications of F-EEM (Coble, 1996, Coble et al., 1993, Her et al., 2003, Matthews et al., 1996). However, there is a need to understand the possible effects of ionic strength on the fluorescence properties of various compounds such as proteins, amino acids and humic substances. Polysaccharides do not fluoresce when excited by light; therefore, they are only "captured" by LC-OCD.

### 2.3.1   EFFECT OF IONIC STRENGTH

Ionic strength was shown to have an important effect on the fluorescence intensity. As it can be observed in Figure 2.2 the measured intensity increases linearly with the increase of salinity represented as TDS. In brackish water (typically TDS < 10 g/L), an increase of 5 to 8 % is observed for amino acids and humic substances while 24 % is observed for a large protein like albumin. In saline environment (TDS = 35 g/L) an increase of 13 to 24 % is observed for amino acids and humic substances.

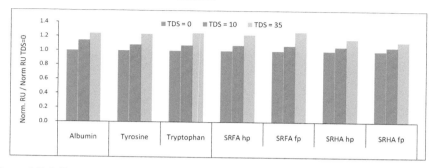

**Figure 2.2. Salinity effect on fluorescence EEM measurements. The intensities are expressed as ratios with respect to the intensity of the sample with TDS = 0 g/L.**

In all cases was observed an increase in intensity with increasing the salinity of the solution. This might be bue to hydrolisis of the organic matter components.

## 2.3.2   EFFECT OF pH

pH of a set of samples with constant concentration of organic matter (OM) was modified by the addition of hydrochloric acid (HCl) or sodium hydroxide (NaOH), and as a result, a variation of the fluorescence intensity was observed (See Figure 2.3). This variation might be attributed to the degradation (hydrolisis) of OM in low and high pH environments.

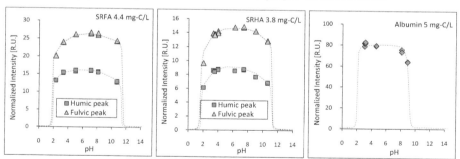

**Figure 2.3. Effect of pH on the measured EEM peak intensities. For humic substances, the intensities at both peaks are shown.**

## 2.3.3   EFFECT OF CONCENTRATION

Analyses of different compound solutions with controlled concentrations show that the observed peak intensities do not always follow a linear trend with the concentration (See Figure 2.4). Albumin shows an acceptable linear trend ($R^2=0.9$). For the other observed compounds, a parabolic or a third order polynomial gave a better fit. For humic substances and as for deep-sea reference material (DSR), both peaks, humic and fulvic, showed a non linear behaviour. It was observed that the intensities for the fulvic peak followed a trend with higher curvature than the intensities for the humic peak.

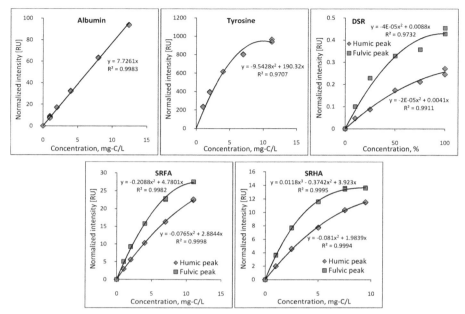

**Figure 2.4. Behaviour of the maximum normalized intensity R.U. with the concentration for different OM compounds. For humic substances, the intensities at both peaks are shown.**

The physical bases of these phenomena are not well understood; however, this effect may be related to the absorption of the emitted light by nearby molecules in the sample. This "obstruction" effect increases with concentration. The use of PARAFAC analysis is based on the assumption of linearity of the response. Thus, linearization of the measurements would be required to obtain accurate results.

## 2.3.4   MAPPING OF STANDARD COMPOUNDS AND REFERENCE MATERIALS

Consensus reference materials (CRM) are available to the international community of dissolved organic carbon (DOC) analysts. The CRM's are used to reference results against the international community of DOC analysts. Deep seawater reference (DSR, Sargasso Sea at 2600 m and Florida Strait at 700 m) are provided by Rosenstiel school of marine and atmospheric science – Division of marine and atmospheric chemistry (RSMAS/MAC) from the University of Miami. In this research, Batch 6 FS – 2006 (Florida Strait at 700 m, 44 - 46 µM DOC and 32.8 µM TN) and Sargasso seawater 0504 (2600 m depth, DOC = 0.54 - 0.56 mg/L, TN = 0.297 µM) were used for "signature" identification of seawater.

These reference materials can be considered without anthropoghenic pollution. Figure 2.5 shows the F-EEM spectra for DSR Florida strait C3 (700 m depth) and DSR Sargasso 0504 (2500 m depth). Both spectra match in the response; they present mainly three regions described by Coble (1996) as humic-like II peak, humic like-I peak and a peak for marine-like humic substances.

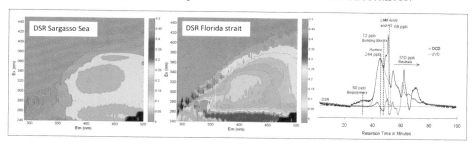

**Figure 2.5. F-EEMs and LC-OCD for DSR Florida strait C3 and DSR Sargasso seawater**

The F-EEM spectrum shows response in areas corresponding typically to humic-like primary and humic-like secondary peaks. In the same way, typical marine humic-like response can be observed. In the LC-OCD chromatogram, the effect of salinity can be observed as a negative response by the DOC and $UV_{254}$ detectors. The humic substances are the main fraction with a concentration of 0.244 mg/L and the biopolymer fraction is not detected by F-EEM but LC-OCD shows a concentration of ~0.050 mg/L.

Figure 2.6 shows the intensities at chosen typical-locations (see Table 2.1) in the F-EEM spectra for Florida strait and Sargasso Sea water reference materials under different conditions of preservation (acidification, pasteurization). Acidified samples DSR FS C3 and DSR Sargasso 0504 matched in their responses.

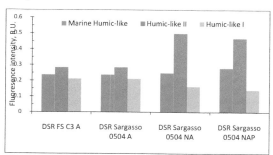

**Figure 2.6. Fluorescence intensity – DSR (acidified (A), not acidified (NA) and not acidified pasteurized (NAP))**

The non acidified sample (DSR Sargasso 0504 NA) produced a higher intensity response for the humic like II peak compared to the acidified samples. In the case of marine humic-like peak, in all cases the response is similar (average 0.251 ± 0.02 stdev). In the case of humic-like I peak the variation is higher (average 0.1820 ± 0.036 stdev). These results suggest that samples should not be altered by acidification or by pasteurization as these will influence the fluorescence intensity of the sample.

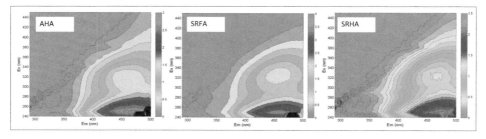

**Figure 2.7. Typical F-EEMs for humic substances: AHA (left), SRFA (middle) and SRHA (right)**

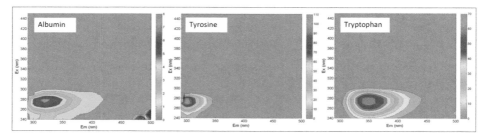

**Figure 2.8. Typical F-EEMs for proteins and amino acids: Albumin (left), Tyrosine (middle) and Tryptophan (right)**

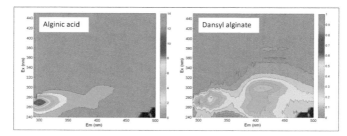

**Figure 2.9. Typical F-EEMs for alginates: Alginic acid (left), Dansyl alginate (right)**

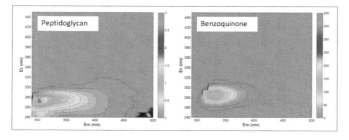

**Figure 2.10. Typical F-EEMs for marine substances: peptidoglycan (left), Quinone (right)**

Figure 2.7 to Figure 2.10 show the fluorescence excitation emission matrices (F-EEMs) for different organic matter substances, and Table 2.3 shows the emission characteristics of these OM compounds. Proteins and amino acids show peak fluorescence emissions of wavelengths between 300-350 nm, and have a single peak: e.g., Albumin, Tyrosine, Tryptophan, Peptidoglycan. Humic substances, e.g., SRFA and SRHA, show peak fluorescence emissions at wavelengths between 300 and 480 nm, and show two peaks: a humic peak (hp) and a fulvic peak (fp). Raman normalized intensities (R.U.) vary largely from one compound to another.

**Table 2.3. Measured Excitation/Emission peak values for standard compounds**

| Description | | Fluorescence range | |
|---|---|---|---|
| | | Ex | Em |
| Terrestrial humic substance | AHA | 250 | 480 |
| | | 310 | 460 |
| Aquatic humic substance | SRHA | 250 | 480 |
| | | 320 | 460 |
| | SRFA | 250 | 450 |
| | | 330 | 460 |
| Protein | Albumin | 280 | 320 |
| Amino acid | Tyrosine | 270-280 | 300 |
| | Tryptophan | 270-280 | 340-360 |
| Alginates | Alginic acid | 270 | 300 |
| | Dansyl alginate | 280 | 320 |
| Peptidoglycan | Peptidoglycan | 270-280 | 300-310 |
| Benzoquinone | 1,4Benzoquinone | 280-300 | 320-340 |
| DSR | Florida strait | 330 | 420 |
| | | 250 | 400-470 |

These F-EEMs and other results from the literature will help to identify the modelled components from PARAFAC analysis.

A comparison of F-EEMs for estuarine water (Amsterdam, DOC ~5 mg/L) and seawater (DOC ~1–1.5 mg/L) is presented in Figure 2.11.

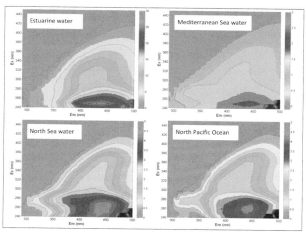

**Figure 2.11. Typical F-EEMs in estuarine water and seawater (Intensities adjusted to actual DOC levels)**

Estuarine water has a 6-10 times higher intensity with respect to seawater due to higher organic matter concentration. Protein-like and humic-like responses are possible to observe

## 2.4   Liquid chromatography – Organic carbon detection

### 2.4.1   EFFECT OF IONIC STRENGTH

The low DOC content and high salinity of seawater, require the adaptation of the standard LC-OCD method already validated for fresh water. Ionic strength was showed to have an important effect in resolution and elution time in the chromatograms as can be observed in Figure 2.12 and Figure 2.13.

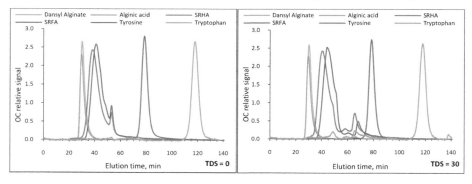

**Figure 2.12. Chromatograms of various standard organic compounds at TDS=0 (left) & TDS=30 (right)**

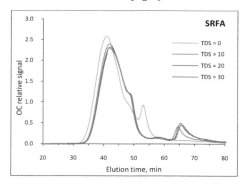

**Figure 2.13. SRFA chromatograms at various TDS**

The high ionic strength of seawater leads to changes in OM configuration and modifies the interaction of the OM with the mobile phase and size exclusion resin. Some of the observed effects when using the standard method developed for fresh water in high salinity water are: *i)* shift in the elution time of the organic matter fractions (longer elution time), this effect was only observed for the humic substances; *ii)* negative depression on the base line at elution times corresponding to acids. The negative peak is probably linked to a strong UV scavenging of salts below 200 nm eluted at this time.

## 2.4.2   TESTING PROTOCOL IN HIGH SALINITY CONDITIONS

The combination of low organic matter concentration and high ionic strength are difficult to resolve in liquid chromatography.

To characterise and quantify organic carbon components in salt and brackish water the approach known as Liquid Chromatography - Organic Carbon Detection (LC-OCD) (Huber, 2007, Huber and Frimmel, 1991, Huber and Frimmel, 1994) was modified to allow the quantification of organic carbon down to the low- concentration range ($\mu$g/L) in salt water matrices.

For optimizing the testing protocol for high salinity waters, several variables in the method were studied, such as: ionic strength of the mobile phase (buffer/eluent), longer column length, oxidant concentrations, acidification rate and dilution of the sample. A more concentrated buffer did not improve the resolution of the chromatogram and the shift in elution time continued. A longer column improved the resolution of the chromatogram and the test increased from 2 hours up to 4 hours.

The modification in the testing protocol for marine waters included: *i)* the prolongation of the chromatographic run by coupling two standard columns in series to improve the chromatogram resolution and *b)* an increase of the flow rate for acidification from 0.3 ml/min to 0.5 ml/min. The latter was required to convert potentially present nanocrystalline carbonate and magnesium/calcium bicarbonate ("Dolomite") to carbonic acid in the inorganic carbon (IC) removal step. These modifications were initially proposed by DOC-LABOR and Gherman and Jacquemet (2007).

The suggested testing conditions for high salinity waters are summarized in Table 2.4.

**Table 2.4. LC-OCD conditions for high salinity waters**

| Parameter | Description |
| --- | --- |
| Column | Toyopearl HW 50S (30 $\mu$m particle size) |
| Number | 2 columns in series |
| Eluent | Same as for fresh water |
| Acidic phase | 0.5 ml/min |
| Dilution factor | 2[1] |
| Software integration | FIFFIKUS[2] |

[1]In particular when testing reverse osmosis concentrates

[2]DOC-LABOR has recently developed a new software for chromatographic integration.

The new conditions allow an acceptable separation of the different organic fractions. The results of a reproducibility test (Table 2.5) showed that parallel samples had variations in the range of $\pm$ 40 $\mu$g/L for DOC concentration and $<$ 20 $\mu$g/L for individual fractions.

**Table 2.5. LC-OCD reproducibility test (Schaule et al., 2010)**

| | DOC | BIO-polymers | DON (Norg) | Humic Subst. (HS) | DON (Norg) | Aromaticity (SUVA-HS) | Mol-Weight (Mn) | Building Blocks | Neutrals | Acids | Inorg. Colloid. SAC |
|---|---|---|---|---|---|---|---|---|---|---|---|
| | | | | Approx. Molecular Weights in g/mol: | | | | | | | |
| | | >>20.000 | | ~1000 (see separate HS-Diagram) | | | 300-500 | <350 | <350 | | |
| | ppb-C | ppb-C | ppb-N | ppb-C | ppb-N | L/(mg*m) | g/mol | ppb-C | ppb-C | ppb-C | (m⁻¹) |
| | % TOC | % TOC | | % TOC | | – | – | % TOC | % TOC | % TOC | |
| Sample 1 | 1474 | 259 | 9 | 435 | 11 | 0,68 | 494 | 162 | 500 | 117 | 0,00 |
| | 100,0 | 17,6 | – | 29,5 | – | – | – | 11,0 | 33,9 | 8,0 | – |
| Sample 2 | 1418 | 249 | 8 | 411 | 11 | 0,95 | 493 | 138 | 507 | 113 | 0,17 |
| | 100,0 | 17,6 | – | 29,0 | – | – | – | 9,7 | 35,7 | 8,0 | – |
| Sample 3 | 1394 | 255 | 9 | 393 | 9 | 1,07 | 495 | 136 | 497 | 112 | 0,13 |
| | 100,0 | 18,3 | – | 28,2 | – | – | – | 9,7 | 35,7 | 8,0 | – |
| Sample 4 | 1470 | 274 | 10 | 406 | 12 | 0,95 | 485 | 144 | 533 | 113 | 0,10 |
| | 100,0 | 18,6 | – | 27,6 | – | – | – | 9,8 | 36,2 | 7,7 | – |
| Sample 5 | 1454 | 272 | 11 | 400 | 15 | 0,94 | 472 | 146 | 524 | 113 | 0,17 |
| | 100,0 | 18,7 | – | 27,5 | – | – | – | 10,0 | 36,1 | 7,7 | – |
| Sample 6 | 1390 | 259 | 12 | 392 | 13 | 0,93 | 474 | 136 | 486 | 117 | 0,19 |
| | 100,0 | 18,6 | – | 28,2 | – | – | – | 9,8 | 35,0 | 8,4 | – |
| average | 1434 | 261 | 10 | 406 | 12 | 0,92 | 486 | 144 | 508 | 114 | 0,13 |
| standard deviation | 38 | 10 | 1 | 16 | 2 | 0,13 | 10 | 10 | 18 | 2 | 0 |
| half confidence interval | | | | | | | | | | | |
| in ppb | 40 | 10 | 1 | 17 | 2 | 0,14 | 11 | 11 | 18 | 3 | 0,07 |
| in % | 2,8 | 3,9 | 14,3 | 4,2 | 18,7 | 14,9 | 2,2 | 7,4 | 3,6 | 2,2 | 57,6 |

## 2.5   Parallel factors analysis

PARAFAC is the acronym of PARallel FACtors. The algorithm was simultaneously developed by Harshman (Harshman, 1970) and Carrol and Chang (Carroll and Chang, 1970) and initially developed for psychometrics and later on applied to chemometrics (Bro, 1997, Geladi, 1989). Using PARAFAC, it is possible under some circumstances, to perform so-called mathematical chromatography; that is, to separate the mixture measurements into the contributions from the underlying individual chemical analyses. For each analysis, the pure excitation and emission spectra are obtained as well as the relative concentration.

The PARAFAC algorithm can be seen as a multi-way extension of PCA. In this study, PARAFAC was used to model the dataset of F-EEMs. It uses an alternating least squares algorithm to minimise the sum of squared residuals in a trilinear model, thus allowing the estimation of the true underlying EEM spectra (Bro, 1997, Bro, 1998). It reduces a dataset of EEMs into a set of trilinear terms and a residual array (Andersen and Bro, 2003). In the three-way case, the algorithm models a three-dimensional array $\underline{X}$. The elements of $\underline{X}$ can be computed the following way by PARAFAC:

$$x_{ijk} = \sum_{f=1}^{F} a_{if}b_{jf}c_{kf} + \varepsilon_{ijk} \qquad\qquad \text{Eq.} \quad 2.1$$

Where $x_{ijk}$ represents an element in $\underline{X}$ in the position given by $i$, $j$ and $k$; $a$, $b$ and $c$ are the loadings and $e_{ikj}$ is the residual (the un-modelled part of the data). In fluorescence spectroscopy, $x_{ijk}$ is the fluorescence intensity of the $i^{th}$

sample at the $k^{th}$ excitation and $j^{th}$ emission wavelength; $a_{if}$ is directly proportional to the concentration of the $f^h$ fluorophore in the $i^{th}$ sample (defined as scores), $b_{jf}$ and $c_{kf}$ are estimates of the emission and excitation spectra respectively for the $f^h$ fluorophore (defined as loadings), F is the number of fluorophores (components) and $\varepsilon_{ijk}$ is the residual element, representing the unexplained variation in the model (Stedmon et al., 2003).

For any data set and choice of parameters, outlying samples must be identified and handled in order for the model to be meaningful. The following approach is taken to identify outliers. For a particular model, samples with a high leverage or high sum-squared residual are removed one by one until no samples are assessed as outliers. Leverages for the samples are defined as the elements on the diagonal of the score matrix.

Some components extracted by PARAFAC can be matched to specific species of organic matter present in water samples, but they more likely represent groups of organic compounds having similar fluorescence properties (Baghoth et al., 2010). While component scores indicate the relative concentrations of groups of organic fractions represented by the components, excitation and emission loadings indicate their characteristic excitation and emission spectra. However, since most of the components that have been extracted from aquatic samples thus far cannot be ascribed to specific organic compounds, the scores cannot be converted to concentrations. Nevertheless, differences in component scores can be used to illustrate variations in the organic matter composition of water samples within a given dataset. Nonetheless, these differences may also be due to changes in the local environment of the analysis, such as polarity and temperature. Differences in scores due to solution environment were minimised by performing fluorescence measurements at the same pH (2.8±0.1) and temperature (20 ± 1° C).

Several diagnostic tools can be used to determine the appropriate number of PARAFAC components. In this study, however, only two methods were mainly employed: split-half analysis (Harshman and Lundy, 1994) and examination of residual error plots (Stedmon and Bro, 2008). For split-half analysis, the dataset was divided into two halves and a PARAFAC model obtained for each half. The excitation and emission spectral loadings of the two halves were then compared to ascertain whether they were similar.

A series of PARAFAC models consisting of between two and seven components were generated using the DOMfluor toolbox (Stedmon and Bro, 2008), which was specifically developed to perform PARAFAC analysis of DOM fluorescence, and contains all of the tools used to identify outlier samples as well as to perform split-half and residual errors diagnostics.

## 2.5.1  PARAFAC RESULTS

A dataset was formed with 124 samples coming from different locations along the Mediterranean and North Sea, all of them corresponding to estuarine/seawater samples before, during and after treatment.

The following methodology was used:

1. Removal of Rayleigh (1st & 2nd) scatter and Raman scatter.
2. Removal of outlier data (visual analysis). A sample was considered an outlier if it contained some instrument error or artifact, or if it was properly measured but was very different from the others (determined by calculating its leverage using DOMfluor). 18 outliers were removed and 106 samples remained in the dataset.
3. Fit models from 2 to n components.
4. Analysis of the reduction of the sum of squared residuals with the number of components.
5. Removal of outliers, i.e., samples with leverage $\geq 0.5$.
6. Repeat steps 3 to 5 while necessary.
7. Perform split half validation. If not successful, go back to step 5.
8. Perform random initial analysis to validate the solution as the optimum.

PARAFAC analysis with 2-7 components was performed on the new dataset. However, only the models containing three, four and six components could be split-half validated. These were split-half validated in the sense that the corresponding components in the split halves had equal excitation and emission loadings as verified by the corresponding Tucker's congruence coefficients being greater than 0.95 (Lorenzo-Seva and Ten Berge, 2006). For a complete dataset model to be validated, the Tucker's congruence coefficients between the split halves, as well as between the complete dataset and a split half should be greater than 0.95 and only the six-component model could be validated in this manner.

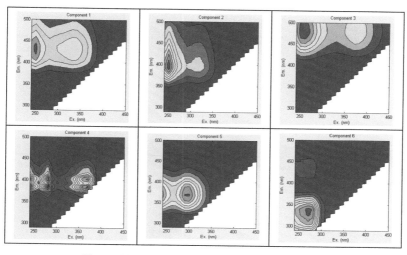

**Figure 2.14. Components from PARAFAC analysis**

Figure 2.14 shows the identified components from PARAFAC analysis and Figure 2.15 shows the loadings versus excitation and emission wavelength pairs of the main peaks of the six components. The description of similar components that were identified in previous studies is presented in Table 2.6.

**Table 2.6. Identification of components**

| Component | Peak(s), (Ex, Em) | Identification |
|---|---|---|
| 1 | (250&330, 440) | SRFA2-like |
| 2 | (250&320, 394) | SRFA1-like |
| 3 | (260&370, 476) | SRHA2-like / AHA-like |
| 4 | (280&375, 282&406) | Marine humic-like? |
| 5 | (300&368) | Marine humic-like |
| 6 | (270&338) | Tryptophan-like |

Comparison of previously identified components with the spectral contours shown in Figure 2.14 and Figure 2.15 indicates that the samples in this study contain humic-like as well as protein-like fluorophores. Components C1 – C5 are humic-like fluorophores. C6 is an amino acid-like fluorophore.

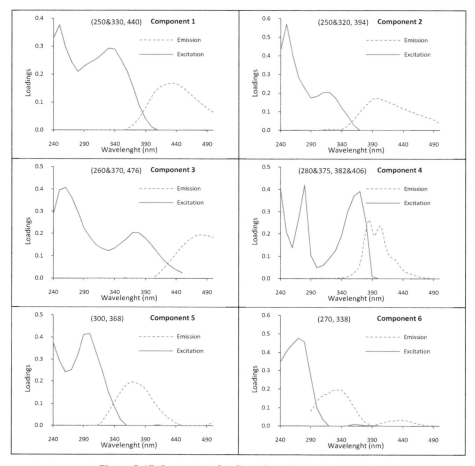

**Figure 2.15. Component loadings from PARAFAC analysis**

PARAFAC analysis was also performed for the standard organic matter compounds described in 2.2.3. In these cases, a dataset was created for each compound together with an artificial component (located at high excitation and low emission wavelengths). The loadings and peak values for excitation and emission wavelengths are presented in the annex.

## 2.6   Principal components analysis

Principal component analysis (PCA) transforms a number of possibly correlated variables into a number of uncorrelated variables called principal components, related to the original variables by an orthogonal transformation. This transformation is defined in such a way that the first principal component has as high a variance as possible (that is, accounts for as much of the variability in the data as possible), and each succeeding component in turn has the highest variance possible under the constraint that it be orthogonal to the preceding components.

There are several methods that can be used to identify factors from a given set of data. The Cattell (1966) scree test and the Kaiser (1960) rule are the most often used procedures to determine the number of components. They are both based on inspection of the correlation matrix eigenvalues. Cattell's recommendation is to retain only those components above the point of inflection on a plot of eigenvalues ordered by diminishing size. Kaiser (1960) recommends that only eigenvalues at least equal to one are retained. One is the average size of the eigenvalues in a full decomposition.

Selection of the appropriate method depends on what one wants to use the analysis for but it is important to consider whether the results will be used to draw conclusions about a population using a sample and whether the data will be used for exploratory purposes or to test a given hypothesis. PCA is one of the methods used to explore data and it assumes that the sample used is the population. In this section, this method was used to explore a set of data comprising LC-OCD results and some operational data such as pre-treatment and origin. All samples correspond to different locations along the Mediterranean and North Sea using reverse osmosis to produce drinking water. Pre-treatment in the plants were: dual media filtration, ultrafiltration, beachwells and in one case an infiltration gallery (or subsurface intake).

## 2.6.1   PCA RESULTS

The software PASW Statistics 18 (from SPSS Inc.) was used for the PCA analysis. The following data was used in the analysis: DOC and LC-OCD fractions (for LMA acids, concentrations were below detection limit in seawater samples) and origin and pre-treatment of the sample if any. The sampling adequacy value KMO (Kaiser-Meyer-Olkin coefficient) was 0.49 from PASW18 analysis. KMO values between 0.5 and 0.7 are considered mediocre but acceptable, while values between 0.7 and 0.8 are good.

The number of factors was automatically generated by the software. Figure 2.16 shows a scree plot of each eigenvalue (Y-axis) against the factor (X-axis) with which it is associated and the plot may be used in selecting the appropriate number of factors to be retained. From the scree plot, it is apparent that two factors are adequate, as was determined by PASW18 using Kaiser's criterion. Since all the communalities, which measures the proportion of variance in a variable, are above 0.7 and the number of variables (5) is less than 30, Kaiser's criterion for retaining factors is sufficient and the three factors extracted by PASW18 were retained for further analysis.

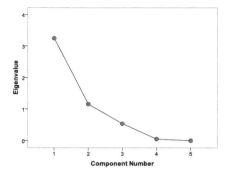

**Figure 2.16. Scree Plot from PCA analysis**

Figure 2.17 shows the component matrix before and after orthogonally rotating the factors. These matrices contain the loadings of each variable onto each factor. After extracting the two components, the next task was to find what these underlying factors (after rotation) represent. Component 1 appears to relate to quantitative measurement of NOM in the water samples; humic substances and neutrals are dominant in the total DOC. Component 2 is a cluster of spectroscopic measurements (biopolymers and building blocks).

| **Component Matrix**[a] | Component | |
|---|---|---|
| | 1 | 2 |
| DOC | .986 | |
| Humic substances | .957 | |
| Neutrals | .946 | |
| Biopolymers | .372 | .807 |
| Building blocks | .573 | .595 |

Extraction Method: Principal Component Analysis.
  a. 2 components extracted.

| **Rotated Component Matrix**[a] | Component | |
|---|---|---|
| | 1 | 2 |
| Humic substances | .981 | |
| Neutrals | .975 | |
| DOC | .965 | |
| Biopolymers | | .888 |
| Building blocks | | .772 |

Extraction Method: Principal Component Analysis.
Rotation Method: Varimax with Kaiser Normalization.
  a. Rotation converged in 3 iterations.

**Figure 2.17. Components matrix from PCA analysis**

In Figure 2.18 the component 1 is plotted versus humic substances and component 2 is plotted vs. the biopolymers concentration in the water samples. The origin of the samples corresponds to: estuarine water (EST), North Sea water (NSW), Mediterranean Sea water (MSW), Pacific Ocean water (POW). A clustering of the samples can be observed according to their origin.

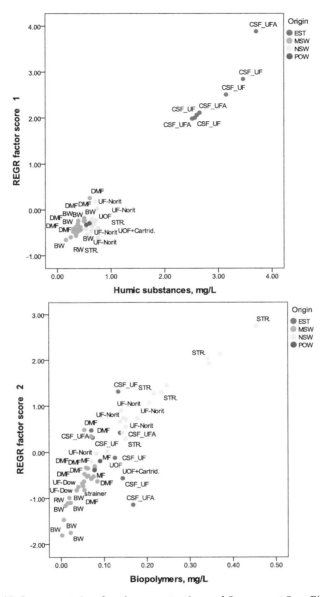

**Figure 2.18. Component 1 vs. humic concentration and Component 2 vs. Biopolymer concentration**

Pre-treatment was classified as: beachwell (BW), dual media filter (DMF), ultrafiltration ~300 kDa (UF-Norit), ultrafiltration ~0.03 μm (UF-DOW), microfiltration ~0.02 μm (MF), strainer 50 μm (STR), coagulation sand filtration (CSF), infiltration gallery (UOF). In all cases, beachwell samples are placed in the lower left indicating the low concentration of humics and biopolymers compared to the other pre-treatments.

## 2.7   Comments

The protocol of fluorescence spectroscopy was tested for ionic strength, pH and organic matter concentration. Higher intensities in the spectra were observed for higher concentrations and high ionic strength. The concentration has a more significant effect than the other two studied variables.

Standard organic matter compounds were mapped by fluorescence spectroscopy. This mapping helped to identify the modelled components by PARAFAC.

In liquid chromatography, high salinity of seawater affects the resolution of the chromatograms, in particular for the humic substances and acids. The testing protocol was modified to improve the resolution of the chromatograms and organic matter recovery.

The PARAFAC model for a dataset with 124 samples produced 6 components. Four of them are related to (marine-) humic substances, one to tryptophan-like organic matter and one possibly due to alginates/humic substances.

The PCA analysis for RO feed water samples that were analysed by LC-OCD showed 2 relevant components. The first components is related to the amount of organic matter present in the water and the second one is related to the amount of biopolymers. Clustering of the samples was observed according to their origin and pre-treatment.

## 2.8   References

AMY, G. & HER, N. (2004). Size exclusion chromatography (SEC) with multiple detectors: a powerful tool in treatment process selection and performance monitoring. *Water science and technology: Water supply*, 4, 19 - 24.

ANDERSEN, C. M. & BRO, R. (2003). Practical aspects of PARAFAC modeling of fluorescence excitation-emission data. *Journal of Chemometrics*, 17, 200-215.

BAGHOTH, S. A., SHARMA, S. K. & AMY, G. L. (2010). Tracking natural organic matter (NOM) in a drinking water treatment plant using fluorescence excitation-emission matrices and PARAFAC. *Water Research*, In Press, Corrected Proof.

BERMAN, T. & PASSOW, U. (2007). Transparent Exopolymer Particles (TEP): An overlooked factor in the process of biofilm formation in aquatic environments. *Nature*. doi:10.1038/npre.2007.1182.1.

BRO, R. (1997). PARAFAC - Tutorial and applications. *Chemometrics and Intelligent Laboratory Systems*, 38, 149-171.

BRO, R. (1998). *Multi-way Analysis in the Food Industry. Models, Algorithms, and Applications.* PhD PhD, University of Copenhagen.

BRO, R. & VIDAL, M. (2010). EEMizer: Automated modeling of fluorescence EEM data. *Chemometr. Intell. Lab. Syst.,* in press.

CARROLL, J. D. & CHANG, I. (1970). Analysis of individual differences in multidimensional scaling via an N-way generalization of and Eckart-Young decomposition. *Psychometrika,* 35, 283.

CATTELL, R. B. (1966). The scree test for the number of factors. *Multivariate Behavioral Research,* 1, 629-637.

COBLE, P. G. (1996). Characterization of marine and terrestrial DOM in seawater using excitation-emission matrix spectroscopy. *Marine Chemistry,* 51, 325-346.

COBLE, P. G., SCHULTZ, C. A. & MOPPER, K. (1993). Fluorescence contouring analysis of DOC intercalibration experiment samples: a comparison of techniques. *Marine Chemistry,* 41, 173-178.

GELADI, P. (1989). Analysis of multi-way (multi-mode) data. *Chemom. Intell. Lab. Syst.,* 7, 11.

GHERMAN, E. & JACQUEMET, V. (2007). Applicability of LC OCD for sea organic matter characterization. Paris: Veolia Environment.

HARSHMAN, R. A. (1970). Foundations of the PARAFAC procedure: Model and conditions for an 'explanatory' multi-mode factor analysis. *UCLA Working Papers in phonetics,* 16, 1-84.

HARSHMAN, R. A. & LUNDY, M. E. (1994). Parafac - Parallel Factor-Analysis. *Computational Statistics & Data Analysis,* 18, 39-72.

HER, N., AMY, G., FOSS, D. & CHO, J. (2002). Variations of molecular weight estimation by HP - size exclusion chromatography with UVA versus on-line DOC detection. *Environmental science and technology,* 36, 3393-3399.

HER, N., AMY, G., MCKNIGHT, D., SOHN, J. & YOON, Y. (2003). Characterisation of DOM as a function of MW by fluorescence EEM and HPLC-SEC using UVA, DOC and fluorescence detection. *Water research,* 37, 4295-4303.

HER, N., AMY, G., PARK, H. & VON-GUNTEN, V. (2004). UV absorbance ratio index with size exclusion chromatography (URI-SEC) as a NOM property indicator

HUBER, S. 2007. *LC-OCD applications* [Online]. DOC-Labor Dr. Huber (online). Available: http://www.doc-labor.de/ [Accessed 01/08/07 2007].

HUBER, S. A. & FRIMMEL, F. H. (1991). Flow Injection Analysis of Organic and Inorganic Carbon in the Low-ppb Range. *Anal. Chem.,* 63, 2122-2130.

HUBER, S. A. & FRIMMEL, F. H. (1994). Direct Gel Chromatographic Characterization and Quantification of Marine Dissolved Organic Carbon Using High-Sensitivity DOC Detection. *Environmental science and technology,* 28, 1194-1197.

KAISER, H. F. (1960). The application of electronic computers to factor analysis. *Educational and Psychological Measurement,* 20, 141-151.

KATSOUFIDOU, K., YIANTSIOS, S. G. & KARABELAS, A. J. (2008). An experimental study of UF membrane fouling by humic acid and sodium alginate solutions: the effect of backwashing on flux recovery. *Desalination,* 220, 214-227.

KLAPPER, L., MCKNIGHT, D. M., FULTON, J. R., BLUNT-HARRIS, E. L., NEVIN, K. P., LOVELEY, D. R. & HATCHER, P. G. (2002). Fulvic acid oxidation state detection using fluorescence spectroscopy. *Environmental science and technology,* 36, 3170-3175.

LEENHEER, J. A., BROWN, P. A. & NOYES, T. I. (1989). *In Aquatic Humic Substances, Influence on Fate and Treatment of Pollutants;,* Washington, DC, American Chemical Society.

LEENHEER, J. A. & CROUÉ, J. P. (2003). Characterizing dissolved aquatic organic matter. *Environmental science and technology,* 18-26.

LEENHEER, J. A., ROSTAD, C. E., GATES, P. M., FURLONG, E. T. & FERRER, I. (2001). Molecular resolution and fragmentation of fulvic acid by electrospray ionization/multistage tandem mass spectrometry. *Anal. Chem.,* 73, 1461-1471.

MATTHEWS, B. J. H., JONES, A. C., THEODOROU, N. K. & TUDHOPE, A. W. (1996). Excitation-emission-matrix fluorescence spectroscopy applied to humic acid bands in coral reefs. *Marine Chemistry,* 55, 317-332.

MCKNIGHT, D. M., BOYER, E. W., WESTERHOFF, P. K., DORAN, P. T., KULBE, T. & ANDERSEN, D. T. (2001). Spectrofluorometric characterization of dissolved organic matter for indication of precursor organic material and aromaticity. *Limnol. Oceanog.,* 46, 38-48.

PICCOLO, A., CONTE, P., TRIVELLONE, E., VAN LAGEN, B. & BUURMAN, P. (2002). Reduced heterogeneity of a lignite humic acid by preparative HPSEC following interaction with an organic acid. Characterization of size-separates by Pyr-GC-MS and 1H-NMR spectroscopy. *Environmental science and technology,* 36, 76-84.

SCHAULE, G., PETROVSKI, K. & HUBER, S. 2010. (Assessment of the biofouling potential of salt and brackish water for RO desalination plants –BDOC/LC-OCD). *In:* IWA (ed.) *MEDINA project.* IWA.

SIGMA-ALDRICH. 2010. *Peptidoglycans (PGN)* [Online]. Sigma-Aldrich. [Accessed].

SIMPSON, A. J., KINGERY, W. L., HAYES, M. H. B., SPRAUL, M., HUMPFER, E., DVORTSAK, P., KERSSEBAUM, R., GODEJOHANN, M. & HOFMANN, M. (2002).

Molecular structures and associations of humic substances in the terrestrial environment. *Naturwissenschaften* 89, 84-88.

SIOUTOPOULOS, D. C., YIANTSIOS, S. G. & KARABELAS, A. J. (2010). Relation between fouling characteristics of RO and UF membranes in experiments with colloidal organic and inorganic species. *Journal of Membrane Science,* 350, 62-82.

SKOOG, D. A. & LEARY, J. J. (1992). *Principles of instrumental analysis,* Orlando, Saunders College Publishing.

STEDMON, C. A. & BRO, R. (2008). Characterizing dissolved organic matter fluorescence with parallel factor analysis: a tutorial. *Limnology and Oceanography - Methods,* 6, 572-579.

STEDMON, C. A., MARKAGER, S. & BRO, R. (2003). Tracing dissolved organic matter in aquatic environments using a new approach to fluorescence spectroscopy. *Marine Chemistry,* 82, 239-254.

VILLACORTE, L. O., KENNEDY, M. D., AMY, G. L. & SCHIPPERS, J. C. (2009a). The fate of Transparent Exopolymer Particles (TEP) in integrated membrane systems: Removal through pretreatment processes and deposition on reverse osmosis membranes. *Water Research,* 43, 5039-5052.

VILLACORTE, L. O., KENNEDY, M. D., AMY, G. L. & SCHIPPERS, J. C. (2009b). Measuring Transparent Exopolymer Particles (TEP) as indicator of the (bio)fouling potential of RO feed water. *Desalination & Water Treatment,* 5, 207-212.

## 2.9    Annex

### 2.9.1    CALCULATION OF THE ACIDS FRACTION WHEN QUANTIFYING NOM

The calculation of the amount of "Acids" in a sample is difficult concerning NOM-analysis. Theoretically arguing is relatively simple, but practical calculation is more difficult.

When NOM-analysis is concerned, in the 4th fraction LMW (Low Molecular Weight) Acids and partly smaller humics are eluting. Possibly charge of the humics is involved in this. Humics or Humic Substances (HS) partly consist of aromatics and will generate an UVD-signal (at 254 nm) next to an OCD-signal. LMW Acids, however, only will display an OCD-signal, because of the lacking of the aromatic structure.

Often, in the chromatogram of the NOM-analysis we see a peak in the 4th fraction at the UVD, which will be attributed to the Humics, not the Acids.

DOC-Labor (provider of LC-OCD and of Fiffikus software) has assumed that the UV/OC ratio (aromaticity) of the main HS-peak must be the same as the smaller Humics in the Acids-peak. They hereby admit that this does not have to be true, but it is a good approximation:

*(UV/OC) main HS-peak = (UV/OC) smaller HS in Acids-peak*

When the OCD-signal of the Acids-peak is relatively too high, concerning the UVD-peak, then after correction the remaining part of the OCD-peak is attributed to the Acids. LMW Acids do not generate an UV-signal namely. Usually however, it is the other way round; when the UVD-signal of the Acids-peak is relatively too high. Then the total amount of the OCD-peak of the Acids is attributed to the Humics (main + smaller parts). This happens in the case of natural waters, in which LMW Acids are biologically unstable and will be decomposed shortly. The part of the UVD-peak, which is relatively too high, is attributed to "SAC inorg. in Acids" in that case. This amount is not visible in the generated report and it means the inorganic UV-absorbing part of the sample (no C-H), that elutes at the place of the Acids, gives no OCD-signal, but a UVD-signal (at 254 nm) indeed. An amount is not calculated, because only amounts will be generated by the OCD-signal. The collected UVD-signal only serves as correction means for the calculation of the Acids, for the visual presentation of the aromatic part of the sample and the calculated aromaticity from this. The part of the Acids-peak of the UVD that is in ratio with the OCD (smaller Humics, no Acids) is called "SAC HS in Acids". Both SAC (Spectral Absorption Coefficient)-values can be found on page 1 of the "Results-file" of Fiffikus.

So, in practice mostly a relatively significant higher UV-signal can be found at the Acids, by which the higher part of the UV-peak is attributed to "SAC inorg. in Acids". At the moment a sample indeed contains LMW Acids, first the UV/OC ratio of the Acids-peak is equalized with the UV/OC ratio of the main Humics-peak, by which a part of the OCD-peak is attributed to the (smaller) Humics, before the remainder is attributed to the Acids. In other words, until the UV/OC ratio's are equal, relative increase of the OC-signal at the Acids-peak is attributed to the Humics first, after which it is attributed to the Acids only.

The calculated amount of Acids always will be too low through this. In this case "SAC inorg. in Acids" always will generate the value "0". This seems to be not true, because in samples without Acids always a significant value is given for "SAC inorg. in Acids". Further investigation is necessary to solve this problem.

The above-mentioned correction of the calculation of the Acids can be made undone, by filling in the value "0" in the "Results-file" of Fiffikus at "tR Humics" on page 1, line 37. The calculation-program then knows that correction of the Acids can be omitted and that the total Acids-peak of the OCD-signal can be attributed to the Acids.

In most of the seawater samples that were analyzed the Acids concentration was below detection limit and therefore is not presented in the following sections.

## 2.9.2 EXCITATION AND EMISSION LOADINGS FOR STANDARD ORGANIC MATTER COMPOUNDS OBTAINED FROM PARAFAC ANALYSIS

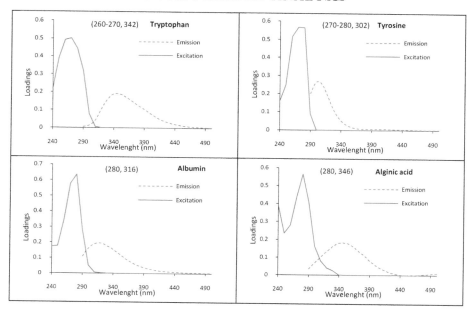

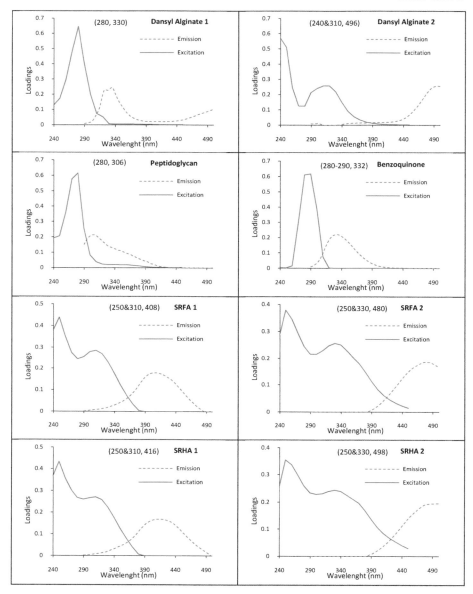

**Figure 2.19. Excitation and emission loadings for standard organic matter compounds obtained from PARAFAC analysis**

# Chapter 3

# 3 Organic foulants in seawater, estuarine and bay sources for reverse osmosis plants

Chapter 3 is based on:

SALINAS RODRÍGUEZ, S. G., KENNEDY, M. D., SCHIPPERS, J. C. & AMY, G. L. (2009). Organic foulants in estuarine and bay sources for seawater reverse osmosis – Comparing pre-treatment processes with respect to foulant reductions. *Desalination and Water Treatment*, 9, 155-164.

BAGHOTH, S. A., MAENG, S. K., SALINAS RODRIGUEZ, S. G., RONTELTAP, M., SHARMA, S., KENNEDY, M. D. & AMY, G. (2008). An Urban Water Cycle Perspective of Natural Organic Matter (NOM): NOM in Drinking Water, Wastewater Effluent, Storm water, and Seawater. *Water Science & Technology: Water Supply*, 8, 701-707.

SALINAS RODRÍGUEZ, S. G., KENNEDY, M. D., SCHIPPERS, J. C. & AMY, G. (2008). Identification of organic foulants in estuarine and seawater reverse osmosis systems – Comparison for different pre-treatments. In: EDS-INSA (ed.) *Membranes in Drinking water production and waste water treatment*. Toulouse, France: European desalination society.

LATTEMANN, S., SALINAS RODRÍGUEZ, S. G., KENNEDY, M., SCHIPPERS, J. C. & AMY, G. (2011). Is seawater desalination green? - An evaluation of state of the art of pre-treatment and desalination technologies considering environmental and performance aspects. In: LIOR, N., BALABAN, M., DARWISH, M., MIYATAKE, O., WANG, S. & WILF, M. (eds.) *Advances in water desalination Vol. 1*. New Jersey: John Wiley & Sons.

# 3.1   Introduction

Membrane fouling in fresh and sea water RO systems is a major operational problem. Not much is known about the role of natural organic matter (NOM) in fouling of RO membranes.

Due to increasing demands an increasing number of countries suffer or will suffer soon from water scarcity. These countries are looking for alternative water sources to satisfy these demands. An attractive alternative for drinking, industrial and agricultural water purposes is the use of estuarine and seawater after treatment by distillation or reverse osmosis. Reverse osmosis is increasingly applied due to lower investment and energy cost. Currently the global production of desalinated water is about 50 $Mm^3$/d and it is projected to double until 2015 (Schiermeier, 2008).

Fouling of membranes in brackish and sea water reverse osmosis is an operational problem. Several types of fouling are observed in practice. This study focuses on organic fouling due to natural organic matter (NOM).

Natural organic matter is a heterogeneous mixture of structurally complex compounds. These compounds are derived from chemical and biological degradation of animals and plants, and NOM is a complex mixture of organic material such as humic substances, hydrophobic acids, carbohydrates, amino acids, carboxylic acids, proteins, hydrocarbons, present in natural fresh water (Croué et al., 1999).

NOM in seawater has mainly been studied by oceanographers, whose main interest is the role of DOC in ocean ecosystems. However, to understand membrane fouling in reverse osmosis (RO) a better characterization of the organic carbon found in estuarine and seawater is required.

Traditionally, the NOM fouling potential of RO feed water has been assessed in terms of DOC, UV absorbance, and colour; however, NOM fouling rates appear not to correlate with these traditional water quality parameters. A problem is that DOC only indicates the amount but not the character of the NOM. More recently, specific UV absorbance (SUVA) has been used to indicate the aromatic character of NOM but SUVA is a direct measure of humic substances which, in relation to non-humic materials, are less problematic as foulants.

This study evaluates the effectiveness of pre-treatment in NOM removal by apllying liquid chromatography with on-line dissolved organic carbon detection (LC-OCD) and, in some cases, fluorescence excitation-emission matrix (F-EEM).

The objective of this chapter is to evaluate the effectiveness of NOM removal by pre-treatment processes used in seawater and estuarine water RO systems.

## 3.2   Material and methods

Focusing on organic (NOM) fouling, the analytical tools that are used in this research include: (i) Liquid chromatography with on-line dissolved organic carbon detection (LC-OCD) and (ii) fluorescence excitation-emission matrix (F-EEM). While there is much experience in applying these techniques to freshwater sources with moderate amounts of DOC, this work evaluates their applicability to seawater with lower amounts of DOC (~0.5 mg/L) and much higher levels of salinity.

### 3.2.1   SOURCE WATERS

Sampling campaigns for the study were performed along the coast of the Mediteranean Sea, North Sea and North Pacific ocean over the period of July 2007 – July 2009. The total number of samples was around 100.

### 3.2.2   LIQUID CHROMATOGRAPHY – ORGANIC CARBON DETECTION

LC-OCD can be used to effectively monitor polar NOM components with a lower SUVA. LC-OCD has been successfully applied to monitoring changes in NOM associated with water treatment (Her et al., 2002). It has also been used to identify problematic NOM components in membrane fouling (Her et al., 2004, Yangali-Quintanilla, 2005). LC-OCD separates NOM according size/molecular weight (MW) classes ranging from higher to lower MW: biopolymers (BP), humic substances (HS), Building blocks (BB), low MW acids (LMA) and neutrals (Ns). The magnitude of the BP peak has been linked with fouling potential in UF membranes (Amy and Her, 2004).

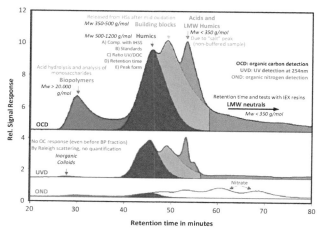

**Figure 3.1. Typical NOM chromatogram of a fresh water sample [Adapted from (Huber, 2007)]**

A typical chromatogram of NOM contained in surface water is shown in Figure 3.1. The first fraction identified after a retention time of approximately

25 - 45 minutes (first peak – largest molecular size) is the biopolymer peak with significant response by organic carbon detection (OCD) only. The organic colloids and proteins present in this fraction provide response in OCD and UV detection. The second and third fraction responses in OCD and UVD are attributed to humic substances and building blocks, respectively. Building blocks are a weathering product of humic substances. The fourth and fifth responses to OCD and UV detection is attributed to low MW acids and neutrals, respectively.

The calculation of the amount of "Acids" in a sample is difficult concerning NOM-analysis. Theoretically arguing is relatively simple, but practical calculation is more difficult. In most of the seawater samples that were analyzed the Acids concentration was below detection limit and therefore is not presented in the following sections.

### 3.2.3   FLUORESCENCE EXCITATION EMISSION MATRIX

Before performing the F-EEM measurement, the DOC of all samples was measured (Shimadzu TOC-V analyzer, Japan) and was adjusted to ~ 1.0 mg/L (as carbon) by diluting samples with milli-Q water having a pH of 2.8 which is the blank sample. Fluorescence EEMs measurement was performed at 240-450 nm (10 nm increments) of excitation wavelength and 290-500 nm (2 nm increments) of emission wavelength using a FluoroMax-3 spectrofluorometer (Horiba Jobin Yvon Inc., USA). The correction steps include a blank subtraction of each EEM, excitation and emission spectra correction using correction factors provided by the manufacturer, and fluorescence intensity normalization with the area of the water Raman peak at excitation wavelength 350 nm.

**Table 3.1. Typical EEM peak values**

| Description | Fluorescence range | |
|---|---|---|
| | Ex | Em |
| Humic-Like Primary Peak | 330–350 | 420–480 |
| Humic-Like Secondary Peak | 250–260 | 380–480 |
| Protein-Like (Tyrosine) Peak | 270–280 | 300–320 |
| Protein-Like (Tryptophan) Peak | 270–280 | 320–350 |
| Protein-Like (Albumin) Peak | 280 | 320 |

Protein-like organic matter, hypothesized to be a principal membrane foulant (polysaccharides are potential foulants as well), exhibits a dominant peak at lower excitation/emission wavelengths while humic/fulvic substances show dominant primary and secondary peaks at higher excitation/emission wavelengths.

### 3.2.4   DEPOSITION FACTOR AND DEPOSITION RATE

Considering the fraction of water that actually passes across the RO membranes, the deposition factor ($\Omega$) is the ratio of the organic matter fraction deposited/accumulated on the RO membrane to that present in that fraction of RO feed water. The relationship between the organic matter fraction concentration of RO concentrate ($C_c$) at recovery $R$ and the organic matter fraction concentration of the RO feed ($C_f$) is used to calculate deposition factor ($\Omega$) as shown in the following equation (Schippers and Kostense, 1980).

$$\Omega = \frac{1}{R} + \frac{C_c}{C_f} \cdot \left(1 - \frac{1}{R}\right)$$

Eq. 3.1

with concentration factor ($CF$):

$$CF = \frac{1 - R \cdot (1 - f)}{1 - R}$$

Eq. 3.2

For $\Omega$ equal to 1 (100 %), the NOM concentrate concentration is equal to the NOM feed concentration, this mean that there is no rejection of NOM by the membranes. For $\Omega$ equal to 0 (0 %), the concentrate concentration is the feed concentration times the concentration factor; in this case NOM is rejected by the membranes. It was assumed that the membranes DOC rejection is 98 %, this is f = 0.98 as this influences the recovery factor. A *positive* deposition factor indicates particles are being deposited as they pass through the system while a *negative* factor indicates the number of particles in the concentrate exceeds the incoming flux (taking into account the concentration factor) (Boerlage, 2001b).

A mass balance for the RO pass was performed as well. From these calculations, the deposition rate in mg/m²-h and the deposition factor were obtained for DOC and for the following NOM fractions: biopolymers, humic substances, building blocks and neutrals. Acids were below detection level in seawater samples.

Particles and organic matter transport in a cross-flow reverse osmosis system involves three process streams: the feed, the permeate and the concentrate. Unlike dead-end filtration, cake formation in cross-flow filtration is limited by back diffusion since most of the rejected particles (and organic matter) remain in suspension flowing towards the concentrate stream. A mass balance was drawn based on this principle to estimate the deposition/accumulation of colloidal particles (and of organic matter) on RO membranes:

$$\frac{dm}{dt} = C_f \cdot Q_f - C_p \cdot Q_p - C_c \cdot Q_c$$

Eq. 3.3

Where, $dm/dt$ is the mass accumulation of particles or organic matter in a period of time. For this, it was assumed that deposition of organic matter or particles is uniform for all RO elements. Subsequently, the deposition rates (DR) in terms of mg-C/m²·h were computed by dividing $dm/dt$ with the total membrane area ($A_m$) of the RO units. Therefore, the deposition rate ($DR$) is:

$$DR = \frac{dm/dt}{A_m} \qquad\qquad\qquad\qquad \text{Eq.} \quad 3.4$$

## 3.3   Results

This section is comprised of two sections. The first deals with characterizing NOM for estuarine and seawater representative samples; while the second deals with the effectiveness of pre-treatment in NOM removal.

### 3.3.1   RAW WATER

Organic matter characterization for representative waters from different locations (estuarine and seawater) and of different organic matter concentrations are presented in this section.

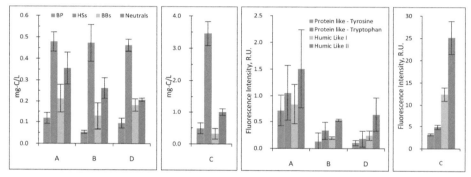

**Figure 3.2. Raw seawater from different locations: LC-OCD (left two) and F-EEM intensities (right two)**

The results of the characterization by F-EEM and by LC-OCD for "raw water" are shown in Figure 3.2. For the "seawater-representative" locations the DOC content is on average 1.08 mg/L where the humic substances represent about 50 % of the DOC content. In all cases the fraction with size larger than 20 kDa (Biopolymers) represents about 7 %. In the case of estuarine water, the DOC content is around 5 mg/L (see Table 3.2).

**Table 3.2. Raw water DOC and SUVA**

| Site | Origin | DOC, mg/L | SUVA, L/mg-m |
|------|--------|-----------|--------------|
| A | Western Mediterranean Sea | 1.16 ± 0.16 | 1.14 ± 0.59 |
| B | Western Mediterranean Sea | 0.92 ± 0.09 | 0.70 ± 0.05 |
| C | Estuary in Amsterdam | 5.26 ± 0.56 | 3.03 ± 0.54 |
| D | Eastern Mediterranean Sea | 0.95 ± 0.03 | 0.89 ± 0.22 |
| G | South Mediterranean Sea | 1.23 ± 0.23 | 1.47 |
| H | South Mediterranean Sea | 1.08 ± 0.18 | 0.80 |
| S | Western Mediterranean Sea | 0.81 ± 0.08 | 1.10 ± 0.58 |
| U | North Pacific Ocean | 1.55 ± 0.1 | 1.55 ± 0.2 |
| Z | North Sea | 1.46 ± 0.18 | 2.05 ± 0.17 |

The samples from site A and site B are from the Western Mediterranean Sea while sample D comes from the Eastern Mediterranean Sea. Site C samples come from an estuary of the North Sea (EC between 1 and 9 mS/cm). The DOC and SUVA values for the plants are shown in Table 3.2. The average values for the other seawater sites (e.g., G, H, S, U and Z) are plotted in Figure 3.3.

Figure 3.3 shows the average values for all analysed samples. Variation represents the maximum and minimum values of the samples. Higher variation is present in humic substances and building blocks in comparison with neutrals and biopolymers.

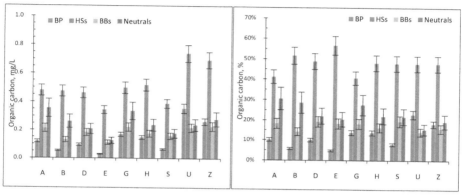

Figure 3.3. Typical organic matter fractions (%) for the samples received (Mediterranean Sea and the North Sea)

LC-OCD and F-EEM results show that humic substances are the more important fraction in seawater and estuarine water. Typically, humic substances (0.5 – 5 kDa in size) represent ~50 % of the DOC content while the biopolymer fraction (> 20 kDa) is less than 8 % of the total DOC. In the case of estuarine water, the humic substances represent ~65 % of the total DOC and biopolymers ~10 %. F-EEM is a good tool to characterize protein-like, humic-like compounds and recent advances with PARAFAC (Parallel factor analysis, a multi-way analysis technique) forecast a major application of fluorescence for water treatment applications as presented by Baghoth et al (2010).

Figure 3.4 shows two graphs related to the nitrogen content in raw water. The figure on the left shows the organic nitrogen concentrations corresponding to biopolymers (DON-BPs) and to humic substances (DON-HSs); the figure on the right shows the percentage of proteins in biopolymers under the presumption that all organic nitrogen in the biopolymer fraction is bound to proteinic matter.

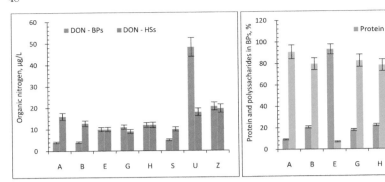

**Figure 3.4. Organic nitrogen concentration for the Biopolymer and Humics fractions (left) and Percentage of protein and polysaccharides in the biopolymer peak (right) for raw seawater**

In general, the organic nitrogen content for biopolymer and for humic substances is less than 0.02 mg/L with exception of site U that presented a nitrogen concentration of around 0.05 mg/L. Regarding the biopolymer fraction of the organic matter, the protein content is less than 20 % with respect to the total concentration of organic carbon in the biopolymer peak.

## 3.3.2   PRE-TREATMENT AND NOM REMOVAL

Three different pre-treatments - beach wells, media filtration, under ocean floor and ultrafiltration - for seawater and in one case estuarine water RO systems were studied, evaluated and compared in terms of organic matter removal and fouling potential.

**Table 3.3. Studied plants/locations**

| Site | Location | Water type | Pre-treatment |
|------|----------|-----------|---------------|
| A | W. Mediterranean Sea | Seawater | Coagulation + Dual media filtration; Microfiltration |
| B | W. Mediterranean Sea | Seawater | Beach well |
| C | North Sea | Estuarine water | Coagulation + Rapid sand filtration + Ultrafiltration |
| D | E. Mediterranean Sea | Seawater | Coagulation + Single stage granular filtration |
| S | W. Mediterranean Sea | Seawater | Ultrafiltration |
| U | North Pacific ocean | Seawater | Infiltration gallery (Subsurface intake) |
| Z | North Sea | Seawater | (Coagulation +) Ultrafiltration |

### 3.3.2.1   Site A – Dual media filtration versus Microfiltration

Site A plant has two parallel treatment trains. The first treatment lane consists of pH correction (6.8 with sulfuric acid), coagulation with iron chloride + polymer addition and dual media filtration (anthracite and sand). The second treatment train consists of pH correction and microfiltration (0.1 μm PVDF membranes operating at 50 L/m$^2$-h).

Table 3.4 shows that the DOC concentration is around 1.2 mg/L at the intake of the plant. SUVA values are in all cases less than 2 L/mg-m, which suggest that NOM is mostly non humics with low hydrophobicity and low molecular weight (Edzwald and Tobiason, 1999).

**Table 3.4. Site A - Raw water DOC and SUVA values**

|                 | Raw water       | Coag + DMF out   | MF out          |
|-----------------|-----------------|------------------|-----------------|
| SUVA, L/mg-m    | $0.78 \pm 0.12$ | $0.69 \pm 0.12$  | $0.59 \pm 0.16$ |
| DOC, mg/L       | $1.19 \pm 0.32$ | $0.77 \pm 0.03$  | $0.85 \pm 0.04$ |

As can be seen in Figure 3.5, the pH correction with sulfuric acid breaks down slightly the humic fraction and the neutrals increase in the same ratio, suggesting possible hydrolysis. In the second train (pH correction and MF) was observed some NOM removal such as humic substances (12 %) and building blocks (25 %).

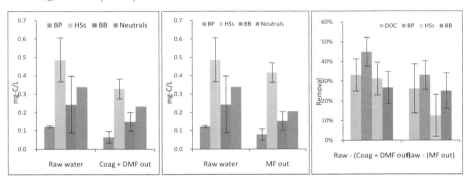

**Figure 3.5. Site A - LC-OCD results: Coag+DMF (left) and Microfiltration (middle) and removal (right)**

Comparing coagulation + DMF and MF can be seen that the former is more effective in removing organic matter, 35 % DOC removal for Coag + DMF compared with 28 % DOC removal for MF. In both treatment trains, the biopolymers are significantly removed (47 % Coag+DMF and 36 % MF). So DMF combined with inline coagulation is more effective than MF without coagulant addition.

### 3.3.2.2    Site B – Beachwells

The beachwells operate at a filtration rate ~0.4 m/h. The DOC content at site B is on average 0.94 mg/L with a SUVA value of around 0.70 L/mg-m. After passage through the beach wells the DOC is reduced to 0.74 mg/L (see Table 3.5).

**Table 3.5. Site B - Raw water DOC and SUVA**

|                 | Raw water       | Beach well      |
|-----------------|-----------------|-----------------|
| SUVA, L/mg-m    | $0.70 \pm 0.05$ | $0.61 \pm 0.28$ |
| DOC, mg/L       | $0.94 \pm 0.09$ | $0.74 \pm 0.06$ |

In Figure 3.6 (left) are the results of LC-OCD analysis. Humic substances are the main fraction in concentration for the raw water and for the beach well effluent.

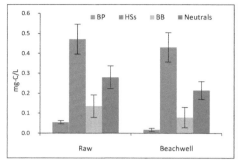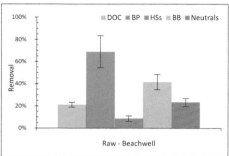

**Figure 3.6. Site B - LC-OCD (left) and NOM Removal (right)**

The NOM removal by the beach wells was around 21 % for DOC, with the biopolymer fraction removed by ~70 %. This is a significant reduction in organic matter with size larger than 20 kDa.

### 3.3.2.3    Site C – Sand filtration and Ultrafiltration – Estuarine water

Inside the site C plant, the water is first dosed with ferric chloride and flows through a continuous sand filter. The filtrate is then fed to the UF system. Normal backwashing of UF is done every 15 minutes and backwashing with NaOCl after 6 hours. A phosphonate based anti-scalant is added to the filtrate before feeding to the RO.

The RO system is composed of Filmtec BW30LE-440 polyamide thin-film composite membranes. Recovery of the plant is about 75 % while salt rejection is 99 %. The RO is chemically cleaned (CIP) thrice a month in summer while only once during non-summer months. See Figure 3.7 for a schematic of the plant.

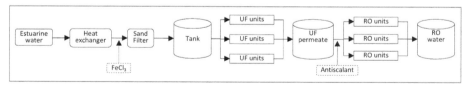

**Figure 3.7. Site C - Scheme of the plant**

On average the raw water DOC for the period considered is around 5 mg/L, turbidity between 5 and 20 NTU, and EC between 1 and 9 mS/cm.

From the LC-OCD results it was observed that the UF backwash produces higher DOC concentration than the UF feed showing its effectiveness (1.25 times higher DOC and 2.9 times higher biopolymers concentration). The RO permeate has a DOC close to 0 mg/L, while the RO concentrate, is 3.75 times the RO feed concentration.

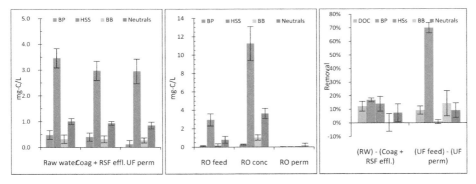

Figure 3.8. Site C - LC-OCD (left and middle) and NOM Removal (right)

From the LC-OCD results it can be observed that humic substances (64 %) are the main component of the water samples. The main biopolymer removal occurs after passing through UF membranes (70 % removal). The RO membranes removed ~95 % DOC of the influent, removing a similar ratio for most of the fractions. The coagulation + sand filtration step removed ~12 % DOC, 17 % biopolymers and 14 % humic substances.

In the case of the UF backwash (not shown) the DOC was found to be 1.22 times more than in the UF feed water; however, there was a significant increase in biopolymers, almost 190 %.

The results of the deposition rate and deposition factor are presented in Table 3.6 corresponding to the plant recovery (R = 75 %). These results suggest that some organic matter deposits on the membranes and large part of the biopolymers are deposited on the membranes. Negative deposition factors for the fractions smaller than 300 Da suggest that these organic fractions may scour from the surface of the membranes.

Table 3.6. Site C - Deposition rate and deposition factor analysis for 5 % deviation in recovery

|  | Deposition rate (mg/m²-h) | | | Deposition factor (f = 0.98) | | |
|---|---|---|---|---|---|---|
|  | 70 % | 75 % | 80 % | 70 % | 75 % | 80 % |
| DOC | -43.9 | 6.7 | 51.0 | -20% | 7% | 30% |
| BP | 1.8 | 2.4 | 3.0 | 51% | 62% | 72% |
| HSs | -27.1 | 8.1 | 39.0 | -22% | 5% | 29% |
| BBs | -6.3 | -2.8 | 0.2 | -55% | -21% | 9% |
| Neutrals | -25.6 | -13.5 | -2.8 | -55% | -21% | 9% |

Deviations may occur in the results due to accuracy of LC-OCD as well as accuracy of the flow meters and readings. These inaccuracies translate in deviations in conversion and ultimately in inaccuracy in deposition factor. To consider these possible inaccuracies on the deposition factor and deposition rate, the effect of 5 % deviation in recovery (see Table 3.6) and the effect of 5

% deviation in feed and concentrate concentrations (see Table 3.7) were calculated.

Table 3.7. Site C - Deposition rate and deposition factor analysis for 5 % deviation in concentrations at R = 75 %

|  | Deposition rate (mg/m²-h) | | | Deposition factor for R = 75 % | | |
|---|---|---|---|---|---|---|
|  | $0.95 \times C_c$, $1.05 \times C_f$ | $C_c$, $C_f$ | $1.05 \times C_c$, $0.95 \times C_f$ | $0.95 \times C_c$, $1.05 \times C_f$ | $C_c$, $C_f$ | $1.05 \times C_c$, $0.95 \times C_f$ |
| DOC | 31.7 | 6.7 | -18.3 | 19% | 7% | -7% |
| BP | 3.0 | 2.4 | 1.9 | 69% | 62% | 55% |
| HSs | 25.4 | 8.1 | -9.1 | 17% | 5% | -9% |
| BBs | -1.4 | -2.8 | -4.3 | -6% | -21% | -37% |
| Neutrals | -8.3 | -13.5 | -18.6 | -6% | -21% | -37% |

The results for the 5 % deviation in concentrations showed for the deposition factor that at least 55 % of biopolymers that are going through the RO system deposited on the membranes with a minimum 1.9 mg/m²-h deposition rate. The results for 5 % deviation in recovery (Table 3.6, R = 70 and 80 %) confirm that biopolymers are deposited on the membranes.

### 3.3.2.4   Site D – Single media filtration

The raw water turbidity ranges from 0.5 to 5 NTU, TOC between 0.7 and 1.5 mg/L, and EC is around 56.5 mS/cm. The pre-treatment of the plant consists of coagulation with ferric sulphate + single stage granular filtration, with the effluent of this step fed to the RO units after cartridge filtration.

The LC-OCD results show that the raw water is mainly composed of humic substances ~50 %, biopolymers ~10 %, and building blocks and neutrals around 20 % each. In terms of organic matter, coagulation + single stage media filtration removes 12 % of DOC where the major removed fraction is the biopolymers (~32 %). The passage of the water through the RO membranes (FILMTEC SW30HR) removes more than 98 % of the organic carbon.

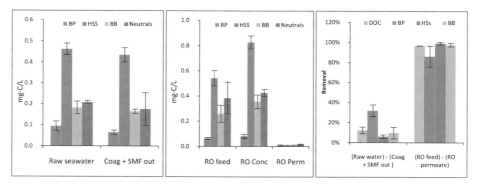

Figure 3.9. Site D - LC-OCD (left) and NOM Removal effectiveness (right)

As explained in section 3.2.4, the deposition rate and the deposition factor were calculated for the first pass of the RO plant in site D. These calculated values are presented in Table 3.8 corresponding to the plant recovery ($R = 48$ %). The results suggest organic matter is accumulating/depositing on the membrane surface. Among the organic matter fractions, the neutrals ($\Omega = 88$ %) have a higher deposition factor compared with biopolymers (65 %), humics (42 %) and building blocks (59 %). DOC accumulated on the membrane surface at a rate of 14.2 mg/m²-h.

It was also projected a possible variation in recovery during operation and variation in organic matter concentration values. The effect of ±5 % deviation in recovery is shown in Table 3.8 and the effect of ±5 % deviation in feed and concentrate concentrations is shown in Table 3.9.

Table 3.8. Site D - Deposition rate and deposition factor including analysis for 5 % deviation in recovery

|  | Deposition rate (mg/m²-h) | | | Deposition factor (f = 0.98) | | |
|  | 43 % | 48 % | 53 % | 43 % | 48 % | 53 % |
|---|---|---|---|---|---|---|
| DOC | 9.8 | 14.2 | 17.8 | 30% | 43% | 53% |
| BP | 0.9 | 1.0 | 1.1 | 58% | 65% | 72% |
| HSs | 5.0 | 7.2 | 9.0 | 29% | 42% | 52% |
| BBs | 4.0 | 4.7 | 5.3 | 50% | 59% | 67% |
| Neutrals | 9.9 | 10.2 | 10.4 | 85% | 88% | 90% |

From Table 3.8 (variation in recovery) can be concluded that a higher recovery increases the accumulation of organic matter while a lower recovery decreases the accumulation of organic matter. In all cases, all the organic matter fractions showed a positive deposition factor and positive deposition rate.

Table 3.9. Site D - Deposition rate and deposition factor analysis for 5 % deviation in concentrations at R = 48 %

|  | Deposition rate (mg/m²-h) | | | Deposition factor for R = 48 % | | |
|  | $0.95 \times C_c$, $1.05 \times C_f$ | $C_c$, $C_f$ | $1.05 \times C_c$, $0.95 \times C_f$ | $0.95 \times C_c$, $1.05 \times C_f$ | $C_c$, $C_f$ | $1.05 \times C_c$, $0.95 \times C_f$ |
|---|---|---|---|---|---|---|
| DOC | 20.7 | 14.2 | 7.7 | 59 % | 43 % | 25 % |
| BP | 1.4 | 1.0 | 0.7 | 79 % | 65 % | 50 % |
| HSs | 10.4 | 7.2 | 4.0 | 58 % | 42 % | 24 % |
| BBs | 6.2 | 4.7 | 3.3 | 74 % | 59 % | 43 % |
| Neutrals | 12.2 | 10.2 | 8.2 | 100 % | 88 % | 75 % |

From Table 3.9 (variation in concentrations) was observed in all cases ($0.95 \times C_c$, $1.05 \times C_f$ & $1.05 \times C_c$, $0.95 \times C_f$), that all the organic matter fractions show a positive deposition factor and positive deposition rate.

A 5 % change in concentrations was more significant that a 5 % change in recovery.

### 3.3.2.5    Site S – Ultrafiltration

The pilot plant (Figure 3.10) consists of ultrafiltration followed by reverse osmosis at 51 % recovery. The pilot plants receive water from an open intake (submerged pipe) located 2.5 km from the coast and 25 m below the surface of the water. The intake pipe is cleaned by chlorination (frequency not disclosed).

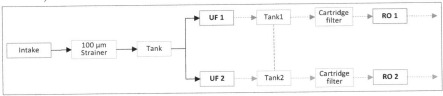

**Figure 3.10. Scheme of the Site S**

Before the raw water being fed to the UF, it passes through an 100 μm strainer. The ultrafiltration units (UF1 and UF2) operate at constant flux at ~60 L/m²-h. Backwash is applied at double the operation flow with air scour every 30 min consisting of 10 seconds air scour, 15 seconds backwash with UF permeate and 45 seconds forward flush with raw water.

**Table 3.10. UF1 and UF2 units' description**

| Parameter | Value | Comment |
|---|---|---|
| Operation | Constant flux (1.9 m³/hr) | (typical pressure 0.7 bar) |
| Flux | ~58 L/m².h | |
| Nominal pore size | 0.03 μm | |
| Material | PVDF | |
| Brand | SFP-2660 | OM Exell - DOW |
| Backwash | 70 sec. | with air scour, permeate water |
| CEB | Every 24 hours | ~ 48 cycles |
| Membrane area | 33 m² | |
| Filtration | Outside to inside | |

The RO systems consists of two units working in parallel (RO1 and RO2). Each unit has six, 4" SW30-4040HR elements and operate at 51 % recovery. Both RO units operate at constant pressure (70 bar). Bi-sulfite and antiscalant are added in front of the RO. The RO production capacity is around 0.76 m³/hr.

The LC-OCD results from the plant are presented in Figure 3.11. The DOC removal achieved by the UF units was around 4 %. Biopolymers were removed by ~14 % on average.

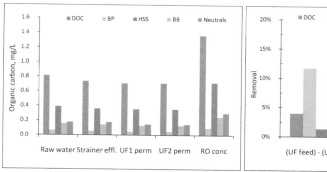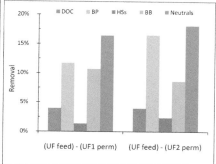

**Figure 3.11. Site S - LC-OCD results (left) and removal (right)**

Figure 3.12 shows the organic nitrogen concentrations for biopolymers and humic substances. The raw water has an average of 5 and 10 µg/L for biopolymers and humic substances, respectively.

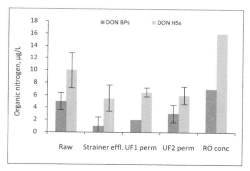

**Figure 3.12. Site S - Organic nitrogen concentration in biopolymer and humics fraction**

The deposition rate and the deposition factor were calculated for the RO units at site S (see Table 3.11) corresponding to the plant recovery (R = 51 %). The results suggest organic matter fractions slightly accumulate on the surface of the membranes. In general, the measured deposition factors were low and less than 27 %.

**Table 3.11. Site S - Deposition rate and deposition factor including analysis for 5 % deviation in recovery**

|          | Deposition rate (mg/m²-h) | | | Deposition factor (f = 0.98) | | |
|----------|------|------|------|-------|------|------|
|          | 46 % | 51 % | 56 % | 46 %  | 51 % | 56 % |
| DOC      | -1.6 | 0.6  | 2.4  | -11 % | 9 %  | 26 % |
| BP       | -0.2 | -0.1 | 0.1  | -32 % | -8 % | 11 % |
| HSs      | -1.3 | -0.1 | 1.0  | -24 % | -1 % | 17 % |
| BBs      | -0.2 | 0.2  | 0.6  | -9 %  | 10 % | 27 % |
| Neutrals | -1.0 | -0.5 | -0.1 | -27 % | -4 % | 15 % |

The effect of 5 % deviation in feed and concentrate concentrations are presented in Table 3.12.

CHAPTER 3

**Table 3.12. Site S - Deposition rate and deposition factor analysis for 5 % deviation in concentrations at R = 40 %**

| | Deposition rate (mg/m²-h) | | | Deposition factor for R = 51 % | | |
|---|---|---|---|---|---|---|
| | $0.95 \times C_c$, $1.05 \times C_f$ | $C_c$, $C_f$ | $1.05 \times C_c$, $0.95 \times C_f$ | $0.95 \times C_c$, $1.05 \times C_f$ | $C_c$, $C_f$ | $1.05 \times C_c$, $0.95 \times C_f$ |
| DOC | 2.7 | 0.6 | -1.5 | 27 % | 9 % | -11 % |
| BP | 0.1 | -0.1 | -0.2 | 11 % | -8 % | -30 % |
| HSs | 1.0 | -0.1 | -1.1 | 18 % | -1 % | -22 % |
| BBs | 0.6 | 0.2 | -0.2 | 28 % | 10 % | -9 % |
| Neutrals | -0.1 | -0.5 | -1.0 | 15 % | -4 % | -26 % |

A 5 % deviation in feed and concentrate concentrations showed not significant change in the estimated deposition factors nor deposition rates. In all cases the biopolymers might accumulate on the membranes by less than 11 %.

The fluorescence-EEM spectra for some points along the treatment plant are presented in Figure 3.13.

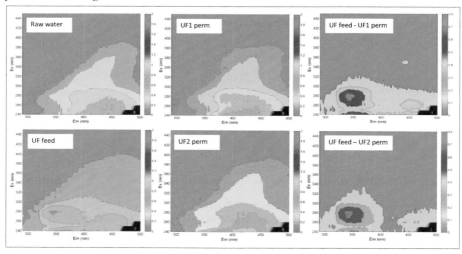

**Figure 3.13. F-EEMs – Raw water, UF1 perm, UF feed-UF1 perm (first row) and UF feed, UF2 perm and UF feed-UF2 perm (second row)**

The fluorescence index for this location is FI = 1.71 which suggests that the organic matter fluorophores are mainly *autochthonous* (microbially-derived). Protein-like (lower left area) and humic-like (lower middle-right and middle right areas) organic matter fluorophores are present in the water samples. According to the differential EEMs, both UF units mainly remove protein-like material.

### 3.3.2.6   Site U – Infiltration gallery (Subsurface intake)

This system (see Figure 3.14 and Figure 3.15) is based on the design criteria associated with slow sand filtration systems. The century-old slow sand filtration concept has been utilized around the world and now offers the opportunity to be applied in an innovative manner for seawater desalination systems. By incorporating slow sand filtration (loading rate of less than 2 $L/m^2$-h) into the seawater collection process, a natural, biological filtration process reduces organic and suspended solids loading on the desalination plant. Therefore, additional pre-treatment is not required, reducing costs, and improving the desalination process. Support gravel layer of around 25 cm and "engineered sand" layer of around 150 cm.

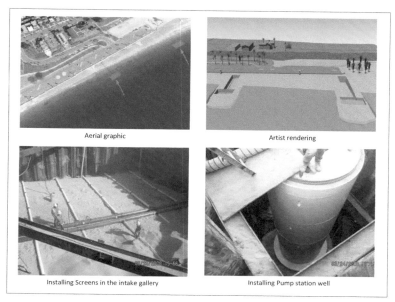

**Figure 3.14. Site U – Infiltration gallery intake & discharge system at Long Beach (Long Beach Water, 2010)**

Some advantages of the infiltration gallery seawater intake system over open ocean intakes or desalination pre-treatment processes are: i) the flow rate and operation of the under ocean floor intake system is unaffected by wave action and tidal forces; ii) it is virtually maintenance free, eliminating operation and maintenance costs; iii) it requires no backwashing, cleaning, treatment, recharging, and/or rehabilitation; and, iv) it serves the dual role of both an intake and pre-treatment component in an environmentally sensitive manner.

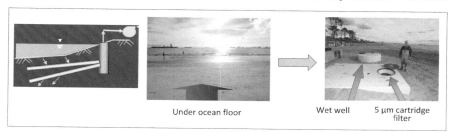

**Figure 3.15. Site U – Infiltration gallery ("Under ocean floor")**

The LC-OCD results obtained from the system are presented in Figure 3.16.

It can be observed that there is a significant removal of biopolymers (~75 %) with this intake system. An additional 5 μm cartridge filter provided around 13 % extra biopolymers reduction.

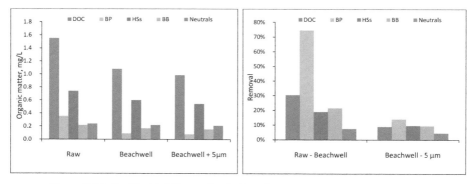

Figure 3.16. Site U – LC-OCD results (left) and removal after beach well (right)

Figure 3.17 shows the organic nitrogen content in the water samples for site U. The most noticeable reduction (~70 %) occurs for the nitrogen in biopolymer fraction after extraction in the beachwell.

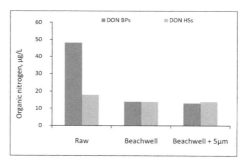

Figure 3.17. Site U - Organic nitrogen in biopolymer and humics fraction

Fluorescence-EEM results are presented in Figure 3.18. Protein-like material and humic-like fluorophores were observed.

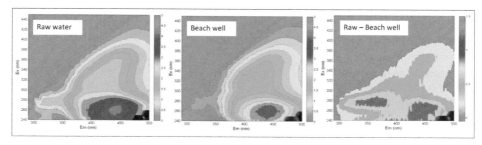

Figure 3.18. Site U - F-EEM for Raw water (left), Beach well (middle) and differential (right)

From Figure 3.18 can be observed that protein like material and humic like material were mainly removed by the under ocean floor (subsurface) intake.

The fluorescence index for the raw water was FI = 1.81 indicating that the organic matter fluorophores are mainly autochthonous (microbially-derived).

### 3.3.2.7    Site Z – Ultrafiltration

The plant consists of a direct intake (L = 100 m) followed by a 50 μm strainer; pH correction and coagulant addition are possible before the UF units (~300 kDa). The UF permeate is stored in a tank that feeds the RO units at a recovery of ~40 % and total capacity of around 15 m³/hr. Average conditions for the UF operation are described in Table 3.13.

**Table 3.13. Site Z - UF unit description**

| Parameter | Value | Comment |
|---|---|---|
| Operation | Constant pressure | It can work at constant flux for short periods |
| Flux | ~60 L/m²-h | After cleaning |
| Nominal pore size | ~300 kDa | |
| Material | PES | |
| Brand | SeaGuard | NORIT filtration |
| Membrane area | 37 m² | Per module |
| Filtration | Inside to outside | |
| Backwash | Every 45 min | |
| Cleaning | 1-2 x /day | |

The results from fluorescence spectrometry for a plant treating water from the North Sea are presented in Figure 3.19. In this case, the fluorescence intensities are presented according to Table 3.1. Humic substances (humic, fulvic and marine-humic like materials) are dominating the spectra with higher fluorescence intensities than the amino-acid like materials.

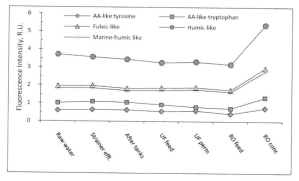

**Figure 3.19. Site Z - Fluorescence intensities (R.U.) along the treatment plant**

From Figure 3.19, the AA-like tyrosine peak was removed by 23 % and the AA-like tryptophan peak was removed by 22 % after the UF units. Coagulation (Strainer effluent – UF feed) removed 15 % AA-like material and 6 %, 9 % and 6 % for fulvic-like, humic-like and marine-humic-like material, respectively. The fluorescence index, FI ~1.75, indicates that the organic matter fluorophores are mainly autochthonous, this is, microbially-derived.

The average LC-OCD results for the sampling campaigns are presented in Figure 3.20. Biopolymers are mainly removed by the UF units (~50 %).

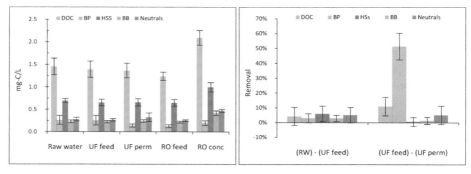

**Figure 3.20. Site Z - Organic matter fractions (left) and removal (right)**

Figure 3.21 shows the organic nitrogen concentrations for biopolymer and humics fraction along the treatment plant. It can be observed that the nitrogen in the biopolymer fraction decreases after UF passage.

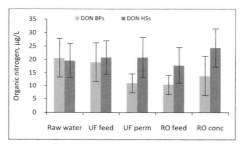

**Figure 3.21. Site Z - Organic nitrogen content for biopolymer and humics**

As explained in section 3.2.4, the deposition rate and the deposition factor were calculated for the RO system in site Z. These calculated values are presented in Table 3.14 corresponding to the system recovery ($R = 40$ %).

The results suggest that organic matter may slightly deposit or accumulate on the membrane surface. Among the organic matter fractions, the humic substances ($\Omega = 10$ %) may accumulate on the membrane surface.

It was also projected a possible variation in recovery during operation and variation in organic matter concentration values. The effect of $\pm 5$ % deviation in recovery is shown in Table 3.14 and the effect of $\pm 5$ % deviation in feed and concentrate concentrations is shown in Table 3.15.

**Table 3.14. Site Z - Deposition rate and deposition factor including analysis for 5 % deviation in recovery**

| | Deposition rate (mg/m²-h) | | | Deposition factor (f = 0.98) | | |
|---|---|---|---|---|---|---|
| | 35% | 40% | 45% | 35 % | 40 % | 45 % |
| DOC | -4.7 | -0.5 | 2.7 | -30 % | -5 % | 14 % |
| BP | -0.4 | 0.0 | 0.3 | -25 % | -1 % | 18 % |
| HSs | -0.8 | 1.0 | 2.5 | -11 % | 10 % | 27 % |
| BBs | -0.9 | -0.1 | 0.4 | -32 % | -6 % | 13 % |
| Neutrals | -0.9 | -0.1 | 0.5 | -29 % | -5 % | 15 % |

From Table 3.14 (variation in recovery) can be observed that a higher recovery increases the accumulation of organic matter while a lower recovery decreases the accumulation of organic matter. At 45 % recovery, all the OM fractions would deposit, while at 35 % recovery, none of the OM fractions would deposit on the RO membrane surface.

**Table 3.15. Site Z - Deposition rate and deposition factor analysis for 5 % deviation in concentrations at R = 40 %**

| | Deposition rate (mg/m²-h) | | | Deposition factor for R = 40 % | | |
|---|---|---|---|---|---|---|
| | $0.95 \times C_c$, $1.05 \times C_f$ | $C_c$, $C_f$ | $1.05 \times C_c$, $0.95 \times C_f$ | $0.95 \times C_c$, $1.05 \times C_f$ | $C_c$, $C_f$ | $1.05 \times C_c$, $0.95 \times C_f$ |
| DOC | 3.7 | -0.5 | -4.8 | 20 % | -5 % | -32 % |
| BP | 0.4 | 0.0 | -0.4 | 23 % | -1 % | -28 % |
| HSs | 3.2 | 1.0 | -1.1 | 33 % | 10 % | -15 % |
| BBs | 0.6 | -0.1 | -0.9 | 18 % | -6 % | -34 % |
| Neutrals | 0.7 | -0.1 | -1.0 | 20 % | -5 % | -32 % |

From Table 3.15 (variation in concentrations) was observed for $0.95 \times C_c$ and $1.05 \times C_f$, that all the organic matter fractions show a positive deposition factor and positive deposition rate. In the opposite case none of the fractions would deposit.

A 5 % change in concentrations was more significant that a 5 % change in recovery.

## 3.4   Comments

Seawater and estuarine water were analytically characterized in terms of organic matter by LC-OCD and in some cases by F-EEM. In the case of seawater, on average DOC of 1.08 mg-C/L, humic substances represent ~65 %, biopolymers ~12 %, and neutrals the remaining 23 %. In case of estuarine water, on average 5.2 mg-C/L, humic substances consisted of ~72 %, biopolymers ~10 %, and neutrals the remaining 18 %.

Table 3.16 shows a summary of the pre-treatment NOM removal achieved in the different studied locations.

Beachwells and the infiltration gallery (subsurface intake) removed almost twice the biopolymers (~70 %) compared conventional pre-treatment and membrane pre-treatment. For site C (estuarine water), coagulation + continous sand filtration removed 12 % DOC and 17 % biopolymers. The UF units removed nearly 70 % of the biopolymers that were fed to the RO membranes.

**Table 3.16. Summary pre-treatments - NOM removal efficiencies**

| Site | Pre-treatment | DOC | Biopolymers | Humics |
|------|---------------|-----|-------------|--------|
| A | Coag + DMF | 35% | 47% | 30% |
| A | MF | 26% | 36% | 8% |
| B | Beachwell | 21% | 70% | 9% |
| C | Coag + RSF + UF | 20% | 75% | 15% |
| D | Coag + SMF | 12% | 32% | 6% |
| S | Ultrafiltration | 4% | 15% | 1% |
| U | Infiltration gallery | 30% | 75% | 19% |
| Z | Ultrafiltration | 8% | 51% | 1% |

The deposition factors and deposition rates revealed that some organic matter was deposited on the RO membranes. After considering deviations in concentrations and deviations in recovery, biopolymers are most likely to be accumulated on the membrane surface. This might be an indication of organic matter fouling.

## 3.5    List of abbreviations and symbols

### 3.5.1    ABBREVIATIONS

| | |
|---|---|
| kDa | Kilo Dalton |
| BB | Building blocks |
| BP | Biopolymers |
| DOC | Dissolved organic carbon |
| DR | Deposition rate |
| EC | Electrical conductivity |
| F-EEM | Fluorescence excitation emission matrix |
| LC-OCD | Liquid chromatography with organic carbon detection |
| LMA | Low molecular weight acids |
| HS | Humic substances |
| MFI-UF | Modified fouling index – ultra filtration |
| MWCO | Molecular weight cut off |
| Ns | Neutrals |
| NOM | Natural organic matter |
| R | Recovery |
| RO | Reverse osmosis |
| SUVA | Specific UV absorbance |
| SWRO | Seawater reverse osmosis |
| UF | Ultra filtration |

### 3.5.2    SYMBOLS

| | |
|---|---|
| $A_m$ | Membrane surface area (m$^2$) |
| $dm/dt$ | Mass accumulation on the surface of RO membranes over a period of time |
| $f$ | salt/organic matter passage in a RO membrane |

| $J$ | Permeate water flux (m$^3$/m$^2$·s) |
| $\Omega$ | Deposition factor |

## 3.6   References

AMY, G. & HER, N. (2004). Size exclusion chromatography (SEC) with multiple detectors: a powerful tool in treatment process selection and performance monitoring. *Water science and technology: Water supply*, 4, 19 - 24.

BAGHOTH, S. A., SHARMA, S. K. & AMY, G. L. (2010). Tracking natural organic matter (NOM) in a drinking water treatment plant using fluorescence excitation-emission matrices and PARAFAC. *Water Research*, In Press, Corrected Proof.

BOERLAGE, S. F. E. (2001). *Scaling and Particulate Fouling in Membrane Filtration Systems*. Ph.D. Ph.D., IHE-Delft / Wageningen University.

CROUÉ, J. P., VIOLLEAU, D., BODAIRE, C. & LEGUBE, B. (1999). Removal of hydrophobic and hydrophilic constituents by anion exchange resins. *Water science and technology*, 40, 207 - 214.

EDZWALD, J. K. & TOBIASON, J. E. (1999). Enhanced Coagulation: US Requirements and a Broader View. *Water Science and Technology*, 40, 63-70.

HER, N., AMY, G., FOSS, D. & CHO, J. (2002). Variations of molecular weight estimation by HP - size exclusion chromatography with UVA versus on-line DOC detection. *Environmental science and technology*, 36, 3393-3399.

HER, N., AMY, G., PARK, H. & VON-GUNTEN, V. (2004). UV absorbance ratio index with size exclusion chromatography (URI-SEC) as a NOM property indicator

HUBER, S. 2007. *LC-OCD applications* [Online]. DOC-Labor Dr. Huber (online). Available: http://www.doc-labor.de/ [Accessed 01/08/07 2007].

LONG BEACH WATER. 2010. *Under Ocean Floor Seawater Intake and Discharge Demonstration System* [Online]. California. [Accessed 12 November 2010].

SCHIERMEIER, Q. (2008). Purification with a pinch of salt. *Nature*, 452, 260-261.

SCHIPPERS, J. C. & KOSTENSE, A. (Year). The effect of pretreatment of River Rhine on fouling of spiral wound reverse osmosis membranes. *In:* Proceedings of the 7th International Symposium on Fresh Water from the sea, 1980 Amsterdam. 297-306.

YANGALI-QUINTANILLA, V. A. (2005). *Colloidal and Non colloidal fouling of UF membranes: Analyses of membrane fouling and cleaning*. M.Sc. Thesis M.Sc., Unesco-IHE.

# Chapter 4

# 4  A review of fouling indices used in RO systems

Chapter 4 is based on:

SALINAS RODRÍGUEZ, S. G., KENNEDY, M. D., AMY, G. & SCHIPPERS, J. C. (2011). A review of fouling indices used in RO systems. *Water Research*, submitted.

## 4.1   Introduction

Reliable methods to predict the fouling potential of RO feed water are important in preventing and diagnosing fouling at the design stage and for monitoring pre-treatment performance during plant operation (Boerlage, 2007b). Particles are one of the possible origins of RO fouling. However, as mentioned by many researchers fouling is complex and it may be due to several contributing factors (Khedr, 2000).

In water treatment systems with membrane technology, it is common practice to judge or assess the quality of a feed/pre-treated water via fouling indices such as silt density index (SDI) and modified fouling index (MFI). Commonly before a RO system, pre-treatment is applied; the purpose is to minimise fouling in any of its forms (particulate, organic, biological, or scaling). A proper pre-treatment is essential for the RO operation as it will increase the lifetime of the RO membrane and will maintain the performance (Fritzmann et al., 2007).

Particulate fouling refers to suspended and colloidal particles present in water. These particles can be: clay minerals, organic materials, coagulants, algae, bacteria as such (not growing), extracellular polymer substances and/or transparent exopolymer particles. Parameters like suspended matter, turbidity and particle counts are unreliable. For this purpose, the SDI is commonly applied as a measure for fouling potential due to particles. Aluminium is measured when, e.g., Alum is used as coagulant. Should preferably be less than 10 µg/L. In general, measuring the concentration of all individual colloidal and suspended particles is very difficult; this is why a "sum parameter" is applied.

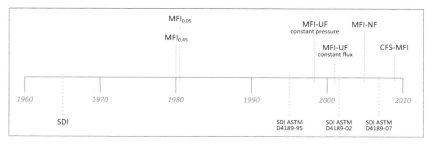

**Figure 4.1. Historical development of fouling indices**

The historical development of fouling indices is presented in Figure 4.1. SDI has a long history in water treatment and it is used worldwide since the 1960s, while the MFI indices are less known though gaining preference in water treatment. All these indices are explained with detail in the following sections.

In this chapter, a review of existent fouling indices such as SDI, $MFI_{0.45}$, MFI-UF constant pressure and MFI-UF constant flux and the cross flow sampler (CFS) coupled with the MFI-UF is presented.

## 4.2   Silt density index

The silt density index (SDI) was introduced by the DuPont company (Permasep Products) at the request of the U.S.A. Bureau of Reclamation. Initially, the test was named the Fouling Index. It was intended to characterize the fouling potential of feed water of DuPont's hollow fine fibre RO permeators (membrane elements). The target contaminants were suspended and colloidal matter. Later on, manufacturers of spiral wound elements and different hollow fibres elements recommended this test as well and formulated maximum levels for SDI to minimize suspended and colloidal fouling and to obtain long-term performance. Currently, SDI < 3 has been set as a requirement for the performance of pre-treatment systems for RO and NF. The SDI is the most commonly used fouling index in water treatment. Additionally, Table 4.1 presents maximum and preferable SDI values according to the membrane type in use.

**Table 4.1. Common SDI guidelines for RO**

| Membrane type | Maximum | Preferable |
|---|---|---|
| Hollow fine fibre (DuPont) | 3 | <1 |
| Hollow fibre (Toyobo) | 4 | |
| Spiral wound | 4-5 | <3 |

Source: Schippers et al. (2010)

The SDI testing procedure is described in the American Society for Testing and Materials (ASTM). The latest version for SDI testing is from 2007 (code 4189-07). The method describes that the SDI test can be used as an indication of the quantity of *particulate* matter (size bigger than 0.45 μm) in water and it should be used for relatively low (<1.0 NTU) turbidity waters such as well water, filtered water, or clarified effluent samples. As the nature of particulate matter in water may vary, the ASTM method indicates that the test is not an absolute measurement of the quantity of particulate matter (ASTM, 2007). Furthermore, it is clearly mentioned that the test is not applicable to permeates from RO and UF systems. This recommendation is not always followed in practice where pre-treatment systems using membrane filtration are assessed with SDI testing. In some cases, high SDI values were obtained after UF pre-treatment that could not be attributed to the "lack of integrity" of the system.

A typical scheme for performing a SDI test is illustrated in Figure 4.2. The $SDI_T$ is calculated from the following equation:

$$SDI_T = \frac{\%PF}{T} = \frac{\left(1 - \frac{t_i}{t_T}\right) \cdot 100}{T}$$          Eq.   4.1

Where, $t_i$ is filtration time of initial filtered volume (min), $t_T$ is the filtration time of second filtered volume (min), $T$ is the total filtration time (min) and $\%PF$ is the percentage of plugging factor. SDI measures the decline in filtration rate expressed in percentage per minute.

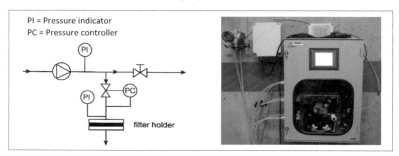

**Figure 4.2. Scheme of an SDI apparatus (left) and picture of an automatic SDI/MFI equipment (right)**

In the test, the sample volume collected is normally 500 ml and SDI 15 minutes ($SDI_{15}$) is the standard SDI. The SDI test is not applicable for all types of water in which the plugging factor (PF) is specified to be not more than 75 % when conducting the test. If this occurs, then the filtration time interval should be reduced to 10 or 5 minutes. If the PF still exceeds 75 % after only 5 minutes of filtration, another procedure should be used to analyze for particulate matter; for example, dilute the sample with water with the same salinity and free from particles. The range of values for the $SDI_5$, $SDI_{10}$ and $SDI_{15}$ are given in Table 4.2.

**Table 4.2. Range values of $SDI_T$ for filtration time T**

| SDI (Filtration time) | $SDI_T$ |
|---|---|
| $SDI_5$ (5 min) | $0 - 20$ |
| $SDI_{10}$ (10 min) | $0 - 10$ |
| $SDI_{15}$ (15 min) | $0 - 6.7$ |

SDI is not based on a fouling mechanism and can never be used to predict the rate of fouling in RO systems where cake filtration is considered the mechanism for particulate fouling. According to Boerlage (2007b), the SDI is based on a mixture of filtration mechanisms; namely blocking (which is not expected for RO membranes) and cake filtration. As the test operates at 2 bar, cake compression will influence the results.

When reporting an SDI value, the following information is required: the SDI with a subscript indicating the total elapsed flow time (T) in minutes; the water temperature before and after the test; and the material and manufacturer of the 0.45 μm membrane filter used for the test.

According to Mosset et al. (2008) and results from other researchers there is no relation between water turbidity and SDI value. In some cases the turbidity values do not change while the SDI values increase.

There are some recommendations for performing a good SDI test: i) the equipment must be flushed before use, ii) purge air to avoid air going at the surface of the membrane, iii) membrane filters must be completely wet, iv) avoid touching the membrane filters with hands.

Some factors affecting the SDI test are described below.

**Membrane**

As mentioned in the ASTM standard and also reported in the literature (Al-hadidi et al., 2008, Al-hadidi et al., 2010, Al-hadidi, 2011, Mosset et al., 2008), the SDI value will vary with: material of the filter, origin of the filter (manufacturer), and even filters in the same production batch. This suggests that SDI values obtained using filters from different membrane manufacturers are not comparable.

Mosset et al. compared SDI values for various hydrophilic membrane materials (nitrocellulose mixed esters, poly-vinylidene fluoride, polytetrafluoroethylene, polyacetylene). Differences of up to 300 % in SDI values were reported. Al-hadidi et al. (2008) reported that there is a variation in membrane properties within a same batch of manufactured membranes (acrylic copolymer, cellulose nitrate, poly-vinylideen-fluoride, poly-tetra-fluoro-ethylene). In his study, the membrane variations were in pore size and roughness up to an average of 10 % and 17 %, respectively, within a batch of membranes, while less variation was observed in bulk porosity which was less than 5 %. The variation in membranes thickness ranged from 3 to 7 % (Al-hadidi et al., 2008). In a study on wastewater reuse, Escobar et al. found a SDI value difference of more than 100 % when using cellulose acetate and nylon membranes (Escobar et al., 2009).

**Temperature**

The viscosity of the water changes with temperature. Cold water has higher resistance to filtration than warm water. For this, any filtration experiment should be normalized to a reference temperature. This is not the case in SDI testing.

**Membrane holder**

Nahrstedt and Camargo (2008) studied the effect of filter support on SDI and MFI values. They reported that the filter holder had a strong influence on the obtained SDI values. A difference of more than 100 % was found for the same feed water depending on the used membrane holder. A similar conclusion was drawn by Escobar et al. (2009) when testing a Millipore holder and a Pall membrane holder.

**Comments:**

Schippers and Verdouw (1980) concluded that the SDI test cannot predict the rate of fouling due to the fact that: i) no linear relation exists between the concentration of suspended and colloidal matter, ii) no correction for temperature, iii) the SDI is not based on any filtration mechanism, iv) it makes use of 0.45 μm filters while pore in RO/NF membrane are approx. 0.001 μm.

It is well known that, even when the recommendations for SDI are not compromised (i.e., SDI < 3 for seawater), serious fouling may occur. This might have two principal reasons: i) other type(s) of fouling occurred and they are not are measured e.g., biofouling, inorganic and organic fouling, fouling due to corrosion products; ii) SDI has no direct predictive value in fouling RO/NF membrane systems. However, it is sometimes an indirect indicator for the fouling potential of RO/NF feed waters.

Furthermore, erratic results are reported with water supersaturated with air; different results are obtained with membranes from different manufacturers; relatively high values are reported in effluents of Micro- and Ultrafiltration systems. The lack of temperature correction and membrane heterogeneity may explain the non uniform results observed in practice.

Despite it being widely used and proven to be of great practical use, Yiantsios et al. (2005) criticised the SDI test as showing no clear correlation between the index value and the fouling behaviour.

A step forward has recently been achieved by normalizing the SDI results based on the clean water flux (membrane resistance). Al-hadidi et al. (2010) proposed a normalization based on the membrane resistance during SDI tests. This correction was proposed initially by Heijman et al (2000) when working with the $MFI_{0.45}$.

Recently, a PhD dissertation has been presented on the limitations, improvements and alternatives of the SDI (Al-hadidi, 2011). Membrane properties of several commercial membranes were shown to influence great deal the measured SDI values. A modified SDI test was also proposed an named as SDI_v in which temperature correction and membrane resistance are considered to minimize erratic results. However, it works under the assumption that complete blocking is the main fouling mechanism in the test and this might not be the case in many waters. A new volume-based SDI was proposed and named as SDI_v in which temperature correction and 0.45 μm filter resistance are considered to minimize erratic results. The SDI_v compares the initial flow rate to the flow rate after filtering a standard volume $V_{f0}$ (~14.58 L). The SDI_v has a linear relationship with the particle concentration assuming cake filtration is dominant during testing. SDI_v shows a more linear relationship to the particle concentration than the standard, time-based SDI.

## 4.3   Modified fouling index

The modified fouling index (MFI) was proposed by Schippers and Verdouw in 1980 from the SDI whereby the same pore size filter (0.45 μm) is used at a constant pressure (2 bar). The $MFI_{0.45}$ considers cake filtration as a fouling mechanism and there is a linear relationship between the MFI and the concentration of foulants in feed water. Cake filtration is considered the dominant fouling mechanism on the surface of a reverse osmosis system (Belfort and Marx, 1979, Schippers and Verdouw, 1980). Furthermore, MFI can be used to predict flux decline or pressure increase to maintain constant capacity in RO systems (Boerlage et al., 2003a, Schippers et al., 1981).

A step forward in order to retain smaller particles was accomplished by using a 0.05 μm membrane filter. The $MFI_{0.05}$ was introduced, and many experiments were carried out with pre-treated river Rhine water to examine the effect on MFI. It was concluded that retained particles larger than 0.45 μm are not responsible for flux decline in RO systems whereas the particles smaller than 0.05 μm, are more likely to be responsible for flux decline (Schippers et al., 1981).

For constant pressure filtration, MFI, independently of the filter pore size, is defined as the slope (tan α) of the graph of $t/V$ vs. $V$ during cake filtration (linear region) and normalized to reference conditions as expressed in Eq. 4.2.

$$MFI = \frac{\eta \cdot I}{2 \cdot P \cdot A^2}$$
Eq.   4.2

The following reference conditions are considered: pressure $(P)$ = 2 bar, membrane area $(A)$ = $13.8 \times 10^{-4}$ m², water temperature through water viscosity $\eta$ at $20°$ C, $V$ = volume in L, $t$ = time in seconds. This definition and conditions were chosen since MFI = 1 s/L² is equivalent to approximately SDI = 1. Therefore, conversion of MFI into $I$ results in $I = 7.6 \times 10^8 \times MFI$.

A different comparison is found in DOW's technical manual that mentions that a MFI value of <1 corresponds to a SDI value of about <3 and this can be considered as sufficiently low to control colloidal and particulate fouling (DOW, 2005). This comparison might come from experimentation but is not mentioned.

To eliminate differences in the $MFI_{0.45}$ due to variation in membrane manufacturer related to pore size and porosity, a reference membrane has been recommended and a correction factor is applied at the beginning of the test based on clean water flux (Heijman et al., 2000).

Measured $MFI_{0.45}$ and SDI values for RO systems feed water are far too low to explain the flux decline rates observed in practice. This suggests that smaller particles are responsible for flux decline rates. To more accurately measure and predict particulate fouling, the MFI has been developed using ultrafiltration (UF) membranes to incorporate fouling due to smaller particles.

It can be calculated using two modes; namely, constant pressure and constant flux (Boerlage, 2001a).

The MFI results depend strongly on the pore size of the filter used. This is illustrated in Figure 4.3 where North Sea water was tested with various membrane pore sizes expressed as MWCO.

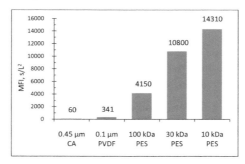

**Figure 4.3. MFI values for North Sea water at various membrane MWCOs. (Constant pressure)**

MFI tests can be performed at constant pressure and at constant flux. At constant pressure (2 bar) the flux decline is monitored while at constant flux the increase in resistance is monitored by the increase in pressure during filtration.

Boerlage et al. (2000, 2003a, 1997, 2002) developed the MFI-UF test to capture smaller particles using a 13 kDa poly-acrylo-nitrile (PAN) UF membrane. Furthermore, the MFI at constant flux was proposed (Boerlage et al., 2003a, Boerlage et al., 2003b).

In an attempt to capture even smaller particles, Khirani et al. (2006b) proposed the MFI-NF at constant pressure with a 0.5 kDa membrane. The MFI-NF is from theoretical point of view impossible because of the occurring phenomenon of concentration polarization (CP) of rejected ions (e.g., mono-valent ions and almost fully divalent ions). The CP will have an dominant effect because of increasing osmotic pressure.

In recent years, several studies have been published dealing with (particulate-) fouling indices. Sim et al. (Sim et al., 2010, Sim et al., 2011a, Sim et al., 2011b) has continued the work of Adham and Fane (2008) on the cross-flow sampler coupled with the MFI-UF test and claimed that the CFS-MFIUF has a lower detection limit than the MFI-UF. Sioutopoulos et al. (2010) has worked in a more applied study with the MFI-UF and able to measure deposition of particles and organic matter in a pilot RO system. Yu et al. (2010) has linked the MFI test with other analytical techniques and defined the particle-MFI, the colloidal-MFI and the organic-MFI. This classification is based on the membrane pore size used in the test.

An example of a filtration set-up to measure MFI-UF at constant pressure is illustrated in Figure 4.4. This set-up is typically observed in lab or bench scale testing. The sample is placed in the sample reservoir (3-5 L). Pressure is controlled with a control valve. Once filtration is started, the weight of the permeate is registered in an electronic balance and recorded in a computer. From the time and volume values, a graph of $t/V$ vs. $V$ is plotted and cake filtration is identified. The slope in this region is the fouling index ($I$).

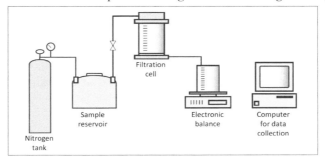

**Figure 4.4. Scheme of filtration setup for MFI measurements at constant pressure**

Figure 4.5 shows the filtration set-up used for MFI-UF constant flux measurements. In this set-up the piston pumps were selected to make sure a pulse-free flow occurs, and in this way a pressure increase development with no artifacts was obtained. The water sample is placed in a syringe that is connected with the membrane holder via a three-way valve. Pressure development in time is monitored. Fouling index is calculated from the linear slope in this curve.

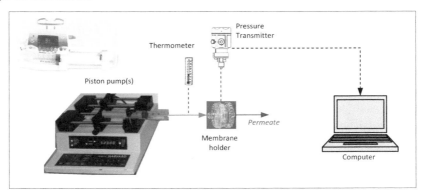

**Figure 4.5. Scheme of filtration setup for MFI measurements at constant flux**

## Comments:

The $MFI_{0.45}$ is widely used in the Dutch drinking water industry and in many desalination plants in France and in Israel to assess the particulate fouling potential of RO feed water and to determine the efficiency of pre-treatment steps. To date, no membrane manufacturer has adopted the MFI as a criterion to assess water quality prior RO systems. However, SDI is not the main criterion anymore.

Constant pressure tests yield very high initial flux rates ($> 1000$ L/m$^2$-h) that are not representative of RO operation (~10-15 L/m$^2$-h). Furthermore, tests at constant pressure are expected to produce different cake properties (porosity) compared to a cake formed on a RO membrane.

## 4.3.1   TEST AT CONSTANT PRESSURE

The MFI test is based on cake filtration, and it can be measured at constant pressure by using the cake filtration equation, where the slope of the linear region in the plot of $t/V$ vs. $V$ was adopted by Schippers and Verdouw (1980) as the MFI. Eq. 4.3 describes the filtration process at constant pressure.

$$\frac{t}{V} = \frac{\eta \cdot R_m}{\Delta P} + \frac{\eta \cdot I}{2 \cdot \Delta P \cdot A^2} \cdot V \qquad\qquad \text{Eq.} \quad 4.3$$

Where, $V$ is the filtrate volume, $t$ is the filtration time, $\Delta P$ is the trans-membrane pressure, $\eta$ is the viscosity of the solution and $A$ is the membrane surface area.

MFI is a function of the dimensions and nature of the particles that form a cake filtration on the surface of the membrane, and is correlated to the concentrations of particles in feed water. The fouling index is normalized to standard reference conditions as mentioned earlier.

The fouling index ($I$) is a product of the specific cake resistance ($\alpha$) and the concentration of particles ($C_b$) in the feed water. The parameters $C_b$ and $\alpha$ are difficult to measure accurately, thus the fouling index ($I$) from filtration test is considered a practical way to measure the fouling potential (Boerlage, 2001a).

$$I = \alpha \cdot C_b \qquad\qquad \text{Eq.} \quad 4.4$$

As such, it is applicable to an incompressible cake. However, if cake filtration is compressible cake compressibility can be taken into account in the fouling index ($I$) equation as follows:

$$I = \alpha \cdot \Delta P^\omega \cdot C_b \qquad\qquad \text{Eq.} \quad 4.5$$

where, $\alpha$ is the specific resistance (constant) and $\omega$ is the compressibility index.

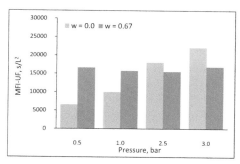

**Figure 4.6. MFI-UF (PES 50 kDa) as function of pressure with and without including compressibility factor**

Figure 4.6 shows MFI at various pressure values for North Sea water. MFI values after including correction for compressibility are also presented. The compressibility coefficient was obtained from the same tests. Tests at high pressure may overestimate the real fouling potential of the water.

In most of the cases, MFI values measured at constant pressure show a pressure dependency due to cake compression indicated by the compressibility coefficient "$\omega$" (e.g., $\omega = 0.82$ for Delft tap water, $\omega = 0.67$ for North seawater). As a consequence, accurate modelling of the rate of membrane fouling is not possible (Boerlage et al., 2004).

The average specific resistance at any applied pressure was found to be always higher during constant pressure filtration than constant flux filtration, except at very low applied pressure. The difference is not significant when the particles are slightly compressible but there is a significant difference in the case where the particles are highly compressible. However, at high constant flux filtration, the resistance becomes closer to that found in constant pressure filtration (Ruth, 1935b). Furthermore, the specific resistance model of cake filtration is based on the assumption that the average cake resistance is constant over time; this implies also that the spatially average values of porosity and pressure differential are also constant over time since specific resistance is a function of both parameters. There have been may attempts to describe this phenomena and to modify the traditional filtration equations (Hieke et al., 2009, Kovalsky et al., 2009, Kovalsky et al., 2007, Tarabara et al., 2002, Theliander and Fathi-Najafi, 1996, Tien and Bai, 2003, Tiller and Cooper, 1960, Tiller and Huang, 1961).

## 4.3.2   TEST AT CONSTANT FLUX

The cakes formed in MFI-UF constant pressure measurements are compressible; consequently, accurate modelling of the rate of fouling RO membranes is not possible. This was the main reason for developing the MFI-UF at constant flux.

Some potential applications of the MFI-UF at constant flux are:

- Assessment of RO and NF feed waters with respect to particulate fouling
- Predicting rate of fouling RO and NF membranes due to particles
- Determining performance of MF/UF systems in particulate removal
- Characterizing MF/UF feed water in predicting development pressure increase during a filtration cycle
- Verifying membrane integrity of MF/UF/RO/NF membrane systems.

In this filtration mode the permeate flux is kept constant and the pressure is increasing to maintain a constant flux. In cake filtration, the pressure increase to keep the flux constant is described in Eq. 4.6.

$$\Delta P = J \cdot \eta \cdot R_m + J^2 \cdot \eta \cdot I \cdot t \qquad\qquad\qquad \text{Eq.} \quad 4.6$$

The fouling index ($I$) can then be determined from the slope of the linear region in a plot of *pressure* vs. *time*, which corresponds to cake filtration. The MFI can be calculated using $I$ (from 4.6) and normalized to standard reference conditions as in Eq. 4.2.

The fouling index can be determined for a shorter time than that calculated with constant pressure (Boerlage et al., 2004).

In principle, the MFI(-UF) test can be performed at any filtration flux. However, at high flux rates two effects may play an important role in the results: flux effect on particle arrangement during cake formation and cake compression during cake formation. These effects are discussed in chapter 6.

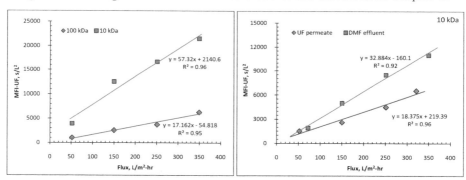

**Figure 4.7. MFI-UF values as function of filtration flux**

Figure 4.7 shows the MFI-UF values for Mediterranean seawater (left) and for UF permeate (0.02 μm pore size) and dual media filtration effluent (right) measured at various flux rates from ~50 L/m²-h up to 350 L/m²-h. From this figure, a direct relation between flux-rate and the measured MFI-UF value was observed. This highly influences the flux rate at which (particulate) fouling indices should operate.

## 4.3.3   THE MFI FOULING PREDICTION MODEL

The MFI models to predict fouling developed by Schippers are based on the assumption that particulate fouling on the surface of reverse osmosis (or nanofiltration) membranes can be described by the cake filtration mechanism (Belfort and Marx, 1979, Schippers et al., 1981). The relationship between the MFI measured for a feed water and the flux decline predicted for a reverse osmosis system are presented below. The relationship is based on the assumption that scaling, adsorptive blocking and biofouling do not contribute to the fouling observed on the RO membrane. Nevertheless, during the MFI test some elements contributing to biofouling might be retained by the membranes (bacteria, organic matter).

RO systems operate in cross flow while the MFI(-UF) is currently a dead-end filtration test. This results in mainly two differences: *i)* in an RO system, not all of the particles are deposited on the surface of the membranes as RO units operate in cross flow, and *ii)* the cake formed in RO has different characteristics than the cake formed in dead-end, e.g., porosity, etc. These differences were respectively translated by Schippers and Kostense (1980) in *i)* the particle deposition factor "$\Omega$" ($\Omega < 1$ for cross flow) and *ii)* the cake ratio factor "$\psi$".

Boerlage et al. (Boerlage et al., 2003a) also made use of this model to predict particulate fouling in freshwater RO systems.

### 4.3.3.1   At constant pressure

The time ($t_r$) in which the flux of a RO membrane has decreased by a factor (e.g., $\Delta J = 15$ %) is:

$$t_r = \frac{\Delta P_r}{\eta_r \cdot J_0^2 \cdot I_r} \cdot \frac{\Delta J \cdot (2 - \Delta J)}{2 \cdot (1 - \Delta J)^2} \qquad \qquad \text{Eq.} \quad 4.7$$

and,

$$I_r = \psi \cdot \Omega \cdot I \qquad \qquad \text{Eq.} \quad 4.8$$

where the subscript $r$ indicates that the parameter refers to filtration through a RO membrane.

Although the proposed model is valid, there are a couple of drawbacks when using the MFI at constant pressure to predict a certain flux decline, namely:

- MFI depends on the pressure, high pressure values will produce high MFI values.
- The pressure drop ($\Delta P_r$) that should be used in the model is not actually known. Unfortunately, there is no possible way to measure the pressure across the formed cake inside the pressure vessel.

### 4.3.3.2    At constant flux

Membrane cleaning is commonly recommended when a 15 % decrease in the normalised flux or increase in pressure drop of an installation is observed.

For a RO system operating under constant flux filtration, the time required for an increase in pressure $\Delta P_r$ to occur can be predicted by:

$$t_r = \frac{(\Delta P_r - \Delta P_{0r})}{J^2 \cdot \eta \cdot I_r}$$

Eq.    4.9

The relationship between $I_r$ and $I$ (from the MFI measurement) was defined in Eq. 4.8 where the cake ratio factor ($\psi$) accounts for differences between the cake deposited on the MFI membrane and that deposited on the RO membrane, and the particle deposition factor ($\Omega$) represents the ratio of the particles deposited on the RO membrane to that present in the feed water.

The particle deposition factor allows to calculate the actual deposition/accumulation of particles in real RO plants and it is specific for each RO plant (flux and recovery) and for each water tested. It is calculated from the relation between the *MFI* of the concentrate at recovery $R$ (of the RO system) and the *MFI* of the feed water as in Eq. 4.10.

$$\Omega = \frac{1}{R} + \frac{MFI_{conc}}{MFI_{feed}} \cdot \left(1 - \frac{1}{R}\right)$$

Eq.    4.10

The prediction model equations (Eq. 4.7 and Eq. 4.9) are a function of the fouling potential of the water at RO operating conditions. The fouling index (I) plays a dominant role as its magnitude depends strongly on the pore size of the filter used. The smaller the filter pore size, the higher the fouling index value and thus shorter estimated time considering a percentage pressure increase.


## 4.4    Cross flow sampler - Modified fouling index

In 2008, the cross flow sampler (CFS) was proposed to simulate the hydrodynamic conditions occurring in a (cross flow) RO process (Adham and Fane, 2008). The hydrodynamic conditions in cross flow filtration mainly refer to the selective deposition of particles. These hydrodynamic conditions in cross flow lead to different cake composition and structure when compared to a dead-end filtration. In order to simulate the selective deposition of particles in a RO system, a cross flow filtration cell with a 5 µm filter was implemented prior to the MFI dead-end cell.

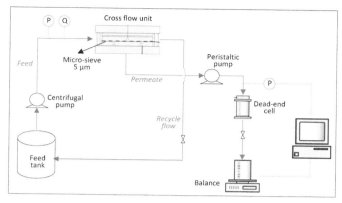

**Figure 4.8. Schematic diagram of CFS-MFI-UF setup**
Based on: Adham and Fane (2008)

CFS-MFI-UF is performed in a typical cross flow filtration unit followed by a dead-end MFI measuring device. A centrifugal pump is used to pump the feed to the cross flow cell and to control the cross flow velocity. A peristaltic pump located at the permeate line feeds the dead-end filtration cell as can be seen in Figure 4.8. Pressure at the dead-end cell is monitored and recorded. A 10 kDa membrane is used in the MFI constant flux test.

Although the idea of considering a cross flow cell to resemble a RO element is remarkable, there are significant differences to consider:

- RO systems operate at recoveries of around 40 % while the cross flow cell only considers velocity of the water. Due to the recovery of the RO system, an increase in ionic strength may influence the particle deposition in full scale RO systems.
- It is possible that by recycling the concentrate, the particle size distribution is modified in the feed water and therefore the deposition of particles will not be characteristic of RO feed water.

Recently, Sim et al. (Sim et al., 2010, Sim et al., 2011a, Sim et al., 2011b) has continued the work of Adham and Fane (2008) on this test and claimed that the CFS-MFI-UF has a lower detection limit than the MFI-UF. Also there has been progress with using this test to predict RO fouling with the cake enhanced osmotic pressure model.

## 4.5 Conclusions

Many studies reported that the main difficulty with SDI, is the lack of reproducible results when performing the tests with various membrane materials and even within the same batch of manufactured filters.

There are two fouling indices namely SDI and $MFI_{0.45}$. For both tests, membranes with pores 0.45 µm are used and measured at constant pressure.

In addition, MFI tests can be done with membranes of different pore sizes down to 5 kDa and at constant pressure and at constant flux.

SDI shows several deficiencies, e.g., no linear relation with concentration of suspended and colloidal matter; no correction for temperature; and it is not based on any filtration mechanism. $MFI_{0.45}$ is a superior alternative since it: shows a linear relation with concentration; is corrected for temperature; and, it is based on cake filtration mechanism.

Both SDI and $MFI_{0.45}$ have no value in predicting the rate of fouling due to particle deposition on RO/NF membrane surfaces. Both might have predictive value in clogging, e.g., non-woven fabric and fibre bundles in DuPont's and Toyobo's permeators and spacers of spiral wound elements.

This is the reason why the MFI-UF – measured with membranes of different pore sizes - has been developed. There are three key issues related with these indices.

-   the pore size or MWCO of the membrane to be used in the test influences greatly the measured values. Furthermore, the MWCO of the membrane should be as close as possible to the pore size of RO membranes if the measured values will be used for fouling prediction.
-   the formation of the fouling layer in the RO system or the deposition / accumulation of particles on the surface of the membranes. In the MFI model, this difference is considered by including the cake ratio factor in the prediction model and in practice is controlled by the flux rate at which filtration occurs.
-   the filtration mode of the MFI test in comparison with the filtration mode of real RO systems (dead-end versus cross flow). This is site specific for each RO plant as it depends on the operational recovery, flux and the water characteristics (particle size distribution in the water). In the MFI prediction model, this is considered by measuring on-site the particle deposition factor in real RO plants.

MFI-UF constant flux has potentially applications in: predicting the rate of fouling on a RO/NF membrane surface due to deposition of particles; verifying performance of MF/UF systems on the removal of colloidal matter; predicting rate of pressure increase in MF/UF systems within a filtration cycle; and verifying membrane integrity of MF/UF/NF/RO membrane systems.

A comparison among SDI, $MFI_{0.45}$, MFI-UF constant pressure, MFI-UF constant flux, and CFS - MFI-UF is presented in Table 4.3. The flux rate during the filtration tests is an important difference. High MFI values will be obtained at high filtration rates. Filtration flux may influence the porosity of the cake by re-arranging the cake structure and compression yielding less porous cakes than those occurring at low flux rates.

The CFS separates hydraulically in a tangential flow the particles smaller than 5 μm. The permeate of the tangential flow filtration is tested in a standard MFI-UF test. Thus, CFS - MFI-UF is not a different test but an application of the MFI-UF to measure the deposition factor in a cell of 20 cm length.

## Table 4.3. Fouling indices comparison

| Parameter | SDI | MFI | MFI-UF cons. pressure[1] | MFI-UF cons. flux[2] | CFS-MFI-UF |
|---|---|---|---|---|---|
| Filtration mode | Dead-end | Dead-end | Dead-end | Dead-end | Dead-end |
| Operation | Constant pressure | Constant pressure | Constant pressure | Constant flux | Constant flux |
| Pore size | 0.45 μm | 0.45 μm | 13 kDa | 10 - 200 kDa | 10 kDa |
| Sample pre-treatment | None | None | None | None | 5 μm cross flow filtration |
| Filter | Flat | Flat | Hollow-fibre | Flat | Flat |
| Fouling mechanism | None | Cake filtration | Cake filtration | Cake filtration | Cake filtration |
| Flux rate | > 1000 L/m²·h | > 1000 L/m²·h | > 1000 L/m²·h | 10 - 350 L/m²·h (⁴) | |
| Membrane properties | Hydrophilic, Mixed CN, M-CA | CA, PVDF, CN | PAN | PES, RC | PES |
| Test recordings | time vs. volume | t/V vs. V (e.g., every 30 s) | t/V vs. V or Δt/ΔV vs. V (e.g., every 10 s) | P vs. t (I vs. t). (e.g., every 10 s). $I = \text{Slope}(P \text{ vs. } t)/(J^2 \cdot \eta)$ | P vs. t |
| Correction/Normalization | None | Temperature, pressure, membrane area | Temperature, pressure, membrane area | Temperature, pressure, membrane area | Same as MFI |
| Formula | $SDI_T = \dfrac{\left(1 - \frac{t_i}{t_T}\right) \cdot 100}{T}$ | $MFI = \dfrac{\eta_{20}}{\eta} \cdot \left(\dfrac{\Delta P}{\Delta P_0}\right)^{1-\omega} \cdot \left(\dfrac{A}{A_0}\right)^2 \cdot \left[\dfrac{d(t/V)}{dV}\right]$ | $MFI = \dfrac{\eta_{20}}{\eta} \cdot \left(\dfrac{\Delta P}{\Delta P_0}\right)^{1-\omega} \cdot \left(\dfrac{A}{A_0}\right)^2 \cdot \left[\dfrac{d(dt/dV)}{2 \cdot dV}\right]$ | $MFI = \dfrac{\eta_{20} \cdot I}{2 \cdot \Delta P_0 \cdot A_0^2}$ | Same as MFI-UF const. flux |
| Cake compression effect | Not considered | ω (compressibility coefficient included in formula. It requires additional testing) | ω = 0 is cake no compressible; ω = 1 is cake very compressible | Controlled by flux rate | - |
| Prediction model | Theoretically impossible | $t_r = \dfrac{\Delta P_r}{\eta_r \cdot J_0^2 \cdot I_r} \cdot \dfrac{\Delta J \cdot (2 - \Delta J)}{2 \cdot (1 - \Delta J)^2}$ | $t_r = \dfrac{\Delta P_r}{\eta_r \cdot J_0^2 \cdot I_r} \cdot \dfrac{\Delta J \cdot (2 - \Delta J)}{2 \cdot (1 - \Delta J)^2}$ | $t_r = \dfrac{(\Delta P_r - \Delta P_{0r})}{J^2 \cdot \eta_r \cdot I_r}$ | - |
| Particle deposition, Ω | Fouling indices accumulate all the particles bigger than filter pore size. | Fouling indices accumulate all the particles bigger than filter pore size. The 0.45μm filter cannot predict real RO fouling rates. | Particles deposition considered through Deposition Factor[3]. | Particles deposition considered through Deposition Factor[3]. | - |

[1] As proposed by Boerlage et al. (2002). [2] As proposed in this dissertation. [3] Chapter 7 discusses deposition of particles in RO systems. [4] MFI test is usually performed at 250 L/m²·h (30 min test) compared with tests at 15 L/m²·h that may last several hours.

## 4.6 Abbreviations and symbols

Abbreviations:

| | |
|---|---|
| CFS | Cross flow sampler |
| M-CA | Mixed cellulose acetate |
| M-CN | Mixed cellulose nitrate |
| MF | Microfiltration |
| MFI | Modified fouling index |
| MFI-UF | Modified fouling index - Ultrafiltration |
| MFI-NF | Modified fouling index - Nanofiltration |
| MWCO | Molecular weight cut-off |
| NDP | Net driving pressure |
| NF | Nanofiltration |
| PAN | Poly-acrylo-nitrile |
| PF | Plugging factor |
| PVDF | Poly-vinylidene fluoride |
| RO | Reverse osmosis |
| SDI | Silt density index |
| SDI_v | Volume-based SDI |
| UF | Ultrafiltration |

Symbols:

| | |
|---|---|
| $A$ | Effective membrane surface area $(m^2)$ |
| $C_b$ | Concentration of particles in a feed water $(kg/m^3)$ |
| $d_p$ | Diameter of particles forming the cake (m) |
| $I$ | Fouling index of particles in water to form a layer with hydraulic resis. $(m^{-2})$ |
| $J$ | Permeate water flux $(m^3/m^2{\cdot}s)$ |
| $R_m$ | Membrane resistance $(m^{-1})$ |
| $t$ | time, (s) |
| $V$ | Filtrate volume $(m^3)$ |
| $\alpha$ | (Average) specific cake resistance (m/kg) |
| $\varepsilon$ | Membrane surface porosity (-) |
| $\eta_T$ | Water viscosity at temperature T $(N{\cdot}s/m^2)$ |
| $\tau$ | Tortuosity of membrane pores |

## 4.7 References

ADHAM, S. & FANE, A. 2008. Cross Flow Sampler Fouling Index. California, USA: National Water Research Institute.

AL-HADIDI, A., KEMPERMAN, A., WESSLING, M. & VAN DER MEER, W. 2008. The influence of membrane properties on the silt density index. *In:* EDS (ed.) *Membranes in drinking and industrial water.* Toulouse: EDS-INSA.

AL-HADIDI, A., KEMPERMAN, A. J. B., SCHIPPERS, J. C., WESSLING, M. & VAN DER MEER, W. G. J. 2010. SDI, is it a reliable fouling index? *In:* EDS (ed.) *Membranes in drinking and industrial water.* Trondheim: EDS/IWA.

AL-HADIDI, A. M. M. 2011. *Limitations, Improvements, Alternatives for the Silt Density Index,* Enschede, Gildeprint Drukkerijen.

ASTM 2007. D4189 - 07 Standard Test Method for Silt Density Index (SDI) of Water. ASTM.

BELFORT, G. & MARX, B. 1979. Artificial particulate fouling of hyperfiltration membranes II. Analyses and protection from fouling. *Desalination,* 28, 13-30.

BOERLAGE, S. F. E. 2001. *Scaling and Particulate Fouling in Membrane Filtration Systems,* Lisse, Swets&Zeitlinger Publishers.

BOERLAGE, S. F. E. 2007. Understanding the SDI and Modified Fouling Indices (MFI0.45 and MFI-UF). *IDA World Congress On Desalination and Water Reuse 2007 - Desalination: Quenching a Thirst* Maspalomas, Gran Canaria - Spain.

BOERLAGE, S. F. E., KENNEDY, M., TARAWNEH, Z., FABER, R. D. & SCHIPPERS, J. C. 2004. Development of the MFI-UF in constant flux filtration. *Desalination,* 161, 103-113.

BOERLAGE, S. F. E., KENNEDY, M. D., ANIYE, M. P., ABOGREAN, E. M., EL-HODALI, D. E. Y., TARAWNEH, Z. S. & SCHIPPERS, J. C. 2000. Modified Fouling Indexultrafiltration to compare pretreatment processes of reverse osmosis feedwater. *Desalination,* 131, 201-214.

BOERLAGE, S. F. E., KENNEDY, M. D., ANIYE, M. P. & SCHIPPERS, J. C. 2003a. Applications of the MFI-UF to measure and predict particulate fouling in RO systems. *Journal of membrane science,* 220, 97-116.

BOERLAGE, S. F. E., KENNEDY, M. D., BONNE, P. A. C., GALJAARD, G. & SCHIPPERS, J. C. 1997. Prediction of flux decline in membrane systems due to particulate fouling. *Desalination,* 113, 231-233.

BOERLAGE, S. F. E., KENNEDY, M. D., DICKSON, M. R., EL-HODALI, D. E. Y. & SCHIPPERS, J. C. 2002. The modified fouling index using ultrafiltration membranes (MFI-UF): characterisation, filtration mechanisms and proposed reference membrane. *Journal of Membrane Science,* 197, 1-21.

BOERLAGE, S. F. E., KENNEDY, M. D., TARAWNEH, Z., ABOGREAN, E. & SCHIPPERS, J. C. 2003b. The MFI-UF as a water quality test and monitor. *Journal of Membrane science,* 211, 271-289.

DOW 2005. *FILMTEC™ Reverse Osmosis Membranes - Technical Manual* DOW.

ESCOBAR, L., SELLERBERG, W., SANCHEZ, D., PASTRANA, F. & WACHINSKI, A. 2009. Detailed analysis of the silt density index (SDI) Results on desalination and wastewater reuse applications for reverse osmosis technology evaluation. *In:* CDA (ed.) *International forum on marine science and technology and economic development - Asia-Pacific Desalination conference*. Qingdao, China: China Desalination Association.

FRITZMANN, C., LÖWENBERG, J., WINTGENS, T. & MELIN, T. 2007. State-of-the-art of reverse osmosis desalination. *Desalination*, 216, 1-76.

HEIJMAN, B., BAAN, E. V. D. & HAAR, G. V. D. 2000. Betrouwbaarheid Membraan-Filtratie-Index kan aanzienlijk vergroot woorden. *H2O*, 12, 21-23.

HIEKE, M., RULAND, J., ANLAUF, H. & NIRSCHL, H. 2009. Analysis of the Porosity of Filter Cakes Obtained by Filtration of Colloidal Suspensions. *Chem. Eng. Technol.*, 32, 1095-1101.

KHEDR, M. G. 2000. Membrane fouling problems in reverse osmosis desalination applications. *Desalination & Water Reuse*, 10, 8-17.

KHIRANI, S., BEN AIM, R. & MANERO, M. H. 2006. Improving the measurement of the Modified Fouling Index using nanofiltration membranes (NF-MFI). *Desalination*, 191, 1-7.

KOVALSKY, P., BUSHELL, G. & WAITE, T. D. 2009. Prediction of transmembrane pressure build-up in constant flux microfiltration of compressible materials in the absence and presence of shear. *Journal of Membrane Science*, 344, 204-210.

KOVALSKY, P., GEDRAT, M., BUSHELL, G. & WAITE, T. D. 2007. Compressible Cake Characterization from Steady-State Filtration Analysis. *AIChE Journal*, 53, 1483-1495.

MOSSET, A., BONNELYE, V., PETRY, M. & SANZ, M. A. 2008. The sensitivity of SDI analysis: from RO feed water to raw water. *Desalination*, 222, 17-23.

NAHRSTEDT, A. & CAMARGO SCHMALE, J. 2008. New insights into SDI and MFI measurements. *Water Science and Technology: Water Supply*, 8, 401-412.

RUTH, B. F. 1935. Studies in Filtration III. Derivation of General Filtration Equations. *Ind. Eng. Chem.*, 27, 708-723.

SCHIPPERS, J. C. 2010. Particulate Fouling. *Membrane Technology in Drinking & Industrial Water Treatment. Principles, Design & Applications*. Delft, The Netherlands: Unesco-IHE.

SCHIPPERS, J. C., FOLMER, H. C. & KOSTENSE, A. Year. The effect of pre-treatment of river rhine water on fouling of spiral wound reverse osmosis membranes. *In:* 7th International Symposium on Fresh water from the Sea, 1980. 297-306.

SCHIPPERS, J. C., HANEMAAYER, J. H., SMOLDERS, C. A. & KOSTENSE, A. 1981. Predicting flux decline or reverse osmosis membranes. *Desalination*, 38, 339-348.

SCHIPPERS, J. C. & VERDOUW, J. 1980. The modified fouling index, a method of determining the fouling characteristics of water. *Desalination* 32, 137-148.

SIM, L. N., YE, Y., CHEN, V. & FANE, A. G. 2010. Crossflow Sampler Modified Fouling Index Ultrafiltration (CFS-MFIUF) – An alternative fouling index. *Journal of Membrane Science,* 360, 174-184.

SIM, L. N., YE, Y., CHEN, V. & FANE, A. G. 2011a. Comparison of MFI-UF constant pressure, MFI-UF constant flux and Crossflow Sampler-Modified Fouling Index Ultrafiltration (CFS-MFIUF). *Water Research,* 45, 1639-1650.

SIM, L. N., YE, Y., CHEN, V. & FANE, A. G. 2011b. Investigations of the coupled effect of cake-enhanced osmotic pressure and colloidal fouling in RO using crossflow sampler-modified fouling index ultrafiltration. *Desalination,* 273, 184-196.

SIOUTOPOULOS, D. C., YIANTSIOS, S. G. & KARABELAS, A. J. 2010. Relation between fouling characteristics of RO and UF membranes in experiments with colloidal organic and inorganic species. *Journal of Membrane Science,* 350, 62-82.

TARABARA, V. V., HOVINGA, R. M. & WIESNER, M. R. 2002. Constant Transmembrane Pressure vs. Constant Permeate Flux: Effect of Particle Size on Crossflow Membrane Filtration. *Environmental engineering science,* 19, 343-355.

THELIANDER, H. & FATHI-NAJAFI, M. 1996. Simulation of the Build-up of a Filter Cake. *Filtration & Separation,* 417-421.

TIEN, C. & BAI, R. 2003. An assessment of the conventional cake filtration theory. *Chemical Engineering Science,* 58, 1323 - 1336.

TILLER, F. M. & COOPER, H. R. 1960. The Role of Porosity in Filtration: IV. Constant Pressure Filtration. *A.I.Ch.E. Journal,* 6, 595-601.

TILLER, F. M. & HUANG, C. J. 1961. Theory. *Filtration equipment,* 53, 529-537.

YIANTSIOS, S. G., SIOUTOPOULOS, D. & KARABELAS, A. J. 2005. Colloidal fouling of RO membranes: an overview of key issues and efforts to develop improved prediction techniques. *Desalination,* 183, 257-272.

YU, Y., LEE, S., HONG, K. & HONG, S. 2010. Evaluation of membrane fouling potential by multiple membrane array system (MMAS): Measurements and applications. *Journal of Membrane Science,* 362, 279-288.

# Chapter 5

# 5 The modified fouling index - ultra filtration - constant flux for seawater applications

Chapter 5 is based on:

SALINAS RODRÍGUEZ, S. G., KENNEDY, M. D., AMY, G. & SCHIPPERS, J. C. (2011). The modified fouling index - ultra filtration - constant flux for seawater applications. *Water Research*, submitted.

SALINAS RODRÍGUEZ, S. G., AL-RABAANI, B., KENNEDY, M. D., SCHIPPERS, J. C. & AMY, G. L. (2009). MFI-UF constant pressure at high ionic strength conditions. *Desalination and Water Treatment*, 10, 64-72.

SALINAS RODRÍGUEZ, S. G., MAMOUN, A., SCHURER, R., KENNEDY, M. D., AMY, G. L. & SCHIPPERS, J. C. (2009). Modified fouling index (MFI-UF) at constant flux for seawater RO applications. In: EDS (ed.) *Desalination for the Environment: Clean water and Energy*. Baden-Baden, Germany: European desalination society.

## 5.1   Introduction

Fouling represents the major constraint to more cost-effective, and therefore expanded, application of membrane technology in drinking water, particularly for reverse osmosis systems. RO membrane modules cannot be pneumatically backwashed and only chemical cleaning can restore normal performance after fouling. Fouling can occur in several forms:

- *Particulate fouling:* small particles and colloids (sub-micron particles) not retained by upstream pre-treatment impart resistance and/or increased salt polarisation and reduce flux as cake-layer deposits accumulate onto the membrane surface.

- *Organic fouling:* natural organic matter (NOM) present in the feed water passing through the pre-treatment processes may be adsorbed onto the membrane surface as a gel-layer, reducing the permeability of the membrane and thereby decreasing production. Moreover, the biodegradable organic matter (BOM) retained on the membrane surface can be utilized by microorganisms as nutrients and may contribute to biological growth.

- *Biofouling:* the combined presence of microorganisms and BOM or nutrients in the membrane feed water may lead to the formation and the development of a biofilm. Microorganisms tend to adhere to surfaces (e.g., membrane surface) and to form a gel layer called biofilm, which participates in the separation process as a secondary membrane. On the raw water side, the biofilm causes an increase of fluid friction resistance which increases the differential feed/concentrate pressure. Also, overall hydraulic resistance of the membrane can increase due to the biofilm. If these effects exceed a certain threshold of interference, they are considered as biofouling (Flemming et al., 1997). The established biofilm promotes further fouling through entrapment of organic molecules, colloidal particles, suspended particles and bacteria cells.

- *Scaling (inorganic fouling):* salt precipitation occurs on the surface of the membrane due to localized supersaturation conditions.

Traditionally, an indirect estimate of particulate fouling potential has been done through the silt density index (SDI) and, more recently, the modified fouling index (MFI) at constant pressure; both rely on the use of a 0.45 μm filter to simulate flux decline trends. The SDI is derived from a simple filtrated volume versus time curve, with a SDI value of less than 3 specified for RO feed water. The MFI, modified from the SDI, has been used to indicate the particulate fouling potential of a feed water. While it is not as widely used, it exhibits more accuracy than the SDI since it is based on a cake filtration mechanism and it is dependent on particle size and particle concentration (Schippers and Verdouw, 1980). In general, smaller particles forming a cake layer result in higher MFI values. As cake layer formation is the dominant mechanism in particulate fouling, the MFI can be used as a basis for modeling flux decline in membrane systems.

A reliable index to predict the fouling potential of RO feedwater is important in preventing and diagnosing fouling at the design stage of RO plants and for monitoring the performance of pre-treatment during plant operation. The silt density index (SDI) is widely used to measure the fouling potential of RO feed water; however, fouling problems have been reported even with very low SDI values, i.e., SDI<1. The modified fouling index (MFI), developed by Schippers and Verdouw (1980), has many advantages over the SDI including: $i)$ a linear relation between the concentration of colloidal particles and MFI $ii)$ cake filtration is assumed to be the dominant filtration mechanism when employing the MFI while the SDI is not based on any filtration mechanism, and $iii)$ since the MFI is based on the occurrence of cake filtration, flux decline in an RO plant can be predicted using cake formation. Both the SDI and MFI operate using constant pressure and a nominal pore size of 0.45 μm. However, particles smaller than 0.45 μm are not captured by the 0.45 μm membrane and thus are not accounted for in the MFI test. It has been reported that particles smaller than 0.45 μm are responsible for fouling of RO membranes. Consequently, the MFI using membranes with a pore size of 0.05 μm was introduced in 1981 (Schippers et al., 1981).

To address the effect of cross flow on the deposition of particles on the membrane surface the deposition factor has been introduced. This factor is the fraction of particles present in the water passing the reverse osmosis or nanofiltration membrane which permanently deposit on the membrane surface (Schippers et al., 1981).

From 1997 to 2001, Boerlage et al. developed the MFI-UF test to capture smaller particles using a polyacrylonitrile (PAN) ultrafiltrafiltration (UF) membrane with a MWCO of 13 kDa. Furthermore, Boerlage et al. also introduced the MFI at constant flux (Boerlage, 2001a, Boerlage et al., 2004, Boerlage et al., 2003a, Boerlage et al., 2003b). In recent times the concept of MFI-NF was proposed to capture even smaller particles, but the measured MFI-NF values for seawater with NF membranes (0.5 kDa) were similar to the MFI-UF values obtained with UF membranes (30 kDa) (Khirani et al., 2006a). The MFI-NF is from theoretical point of view impossible because of the occurring phenomenon of concentration polarization (CP) of rejected ions (e.g., mono-valent ions and almost fully divalent ions). The CP will have an dominant effect because of increasing osmotic pressure.

In 2008, the cross flow sampler (CFS) coupled with the modified fouling index was introduced to consider the hydrodynamic conditions occurring in a (cross flow) RO process (Adham and Fane, 2008). The idea of the process is noteworthy; however, it is not possible to simulate a pressure vessel with a flat sheet membrane unit.

Since the introduction of the MFI-UF by Boerlage et al. (2001a, 2003a), applications have mainly been limited to fresh water sources and the MFI-UF has not yet been tested and evaluated for seawater. Furthermore, MFI-UF constant flux has the main advantage that allows the prediction of rate of fouling in nanofiltration and reverse osmosis systems.

## 5.2    Goal and objectives

The goal of this study is:

- To further develop the modified fouling index constant flux for seawater applications.

The objectives of this chapter are the following:

- To describe the MFI-UF constant flux set-up.
- To characterize the membranes used in the test.
- To investigate variables affecting the MFI-UF tests such as membrane pore size, membrane material and flux rate.
- To apply the MFI-UF test in seawater, in particle size distribution, and in plant profiling.
- To apply the MFI-UF constant flux test to predict pressure development in UF systems and RO particulate fouling prediction.

## 5.3    Material and methods

### 5.3.1   FILTRATION SET-UP

A filtration set-up was developed to work at constant flux as illustrated in Figure 5.1. The key components of the set-up are: pump, membrane holder, pressure sensor, thermometer, computer and a three-way valve.

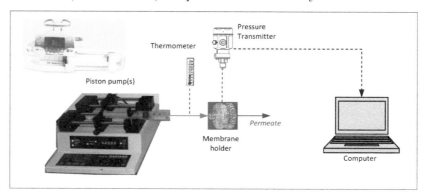

**Figure 5.1. Constant flux filtration set-up**

The set-up was verified for correct pressure readings, constant flux, no leakages, and air trapped in the system.

The readings from the pressure sensor were verified with a second manometer (pressure gauge) by filtering ultra-pure water (UPW) at various flux rates. A maximum 3 % difference was observed, which is considered acceptable.

The flux rate was verified by monitoring the permeate weight over time with help of an electronic balance. Several flux rates (10 - 400 L/m²-h) were tested and a maximum difference of 2.8 % was observed between the expected and measured flux with the lower flux rate.

Verification of leakages in the set-up was performed by pressurizing the system (up to 4 bar) without allowing filtration and monitoring the pressure change over time. No leaks were observed at pressures less than 4 bars over time. However, after stopping the pump a back pulse was observed in the piston pump that yielded a slight decrease in pressure (0.1 pressure loss over 40 min).

The presence of air in the system is not desirable. To verify the effect of air trapped in the system, air was intentionally introduced and filtration was allowed. Erroneous high pressure values were observed by the effect of air; this could be related to the bubble point of the membranes or related to the compression of air that will produce erratic pressure development.

The parts of the filtration set-up are discussed in the following sections.

### 5.3.1.1   Membranes

Two membrane materials with various pore sizes were investigated. The materials were poly ether sulfone (PES) and regenerated cellulose (RC) from Millipore. Both membrane filters are circular flat sheets (25 mm diameter, 0.0004909 m²). The average pressure to filter UPW and the nominal membrane molecular weight cut-off (MWCO) as rated by the manufacturer are summarized in Table 5.1. All membranes tested were new. The stable pressure to filter UPW was measured at 100 L/m²-h in the MFI-UF equipment, then corrected to 20° C.

**Table 5.1. Specifications of the ultrafiltration membranes**

| Material | MWCO, kDa | Clean water pressure (bar) at 20 °C and 100 L/m²-h. |
|---|---|---|
| PES | 5 | 3.4 |
| | 10 | 0.29 |
| | 30 | 0.23 |
| | 50 | 0.18 |
| | 100 | 0.09 |
| RC | 10 | 4.29 |
| | 30 | 0.51 |
| | 100 | 0.14 |

Membrane filters were acquired in packages containing 10 specimens. Each package is numbered with a batch code. For preservation purposes, the manufacturer coats the membranes with glycerine and sodium azide; this coating needs to be removed before the filtration test is performed.

Before testing, all the membranes were soaked in ultra pure water to make sure the membrane is wet and to clean the coating from the manufacturer.

### 5.3.1.2    Constant flow pump

Two piston pumps were tested (shown in Figure 4.5). The maximum pressure at which they can operate is the main difference. Pump 1 (the small one) has a maximum capacity of 1.2 bar while pump 2 (the bigger one) has a capacity of 3.5 bar. This makes pump 1 suitable for working at low flux rates while pump 2 can work at any flux rate and with any membrane pore size.

Some of the features of pump 1 are:

- Flow rate range: 0.1- 200 ml/h, (0.1 ml/h increments)
- Flow rate accuracy: $\pm$ 2 % on syringes
- Pressure: 1.2 bar (max)
- Dimensions (H/W/D/) / Weight: 160 x 345 x 135 mm / 2,150 g
- Battery: Ni/Metal hybrid
- Battery capacity minimum: 10 hours at 5 ml/h per battery
- Brand: Fresenius

The characteristics of pump 2 are:

- Flow rate range: 0.0073 µL/hr - 53 ml/min
- Flow rate accuracy: $\pm$ 3.5 % on syringes
- Force: 40 lbs (177 N)
- Pressure: 3.8 bar (with 60 ml syringe)
- Dimensions (H/W/D/) / Weight: 286 x 311 x 152 mm / 6,800 g
- Battery: None
- Brand: Harvard Apparatus

### 5.3.1.3    Pressure sensor/transmitter

The chosen pressure transmitter is commercially available (Cerabar M HART PMC41, Endress & Hauser) and especially suitable to work with seawater. The operational pressure range is 0-4 bar with a maximum deviation of 0.036 % as illustrated in annex 5.11.3. A three-way valve is used to connect the syringe (water sample) with the membrane holder and at the same time with the pressure transmitter.

The pressure sensor/transmitter has the function to measure and transmit pressure values over time while filtration occurs. The ceramic sensor is illustrated in Figure 5.2 and it consists of: 1) air pressure (gauge pressure sensors), 2) ceramic carrier, 3) electrodes and 4) ceramic diaphragm.

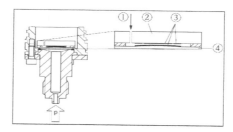

**Figure 5.2. Scheme of the Cerabar M HART pressure transmitter**

The ceramic sensor is a dry sensor with the process pressure acting directly on the rugged ceramic diaphragm and deflecting it a maximum of 0.025 mm. A pressure-proportional change in the capacitance is measured by the electrodes on the ceramic substrate and diaphragm.

Some of the advantages of this pressure transmitter are:

- Maximum deviation in pressure output value is 0.036 % at maximum capacity (~1.5 mbar).
- Guaranteed overload resistance up to 40 times the nominal pressure (max. 60 bar).
- Highly-pure 99.9 % ceramic (Ceraphire®)
- Extremely high chemical stability (corrosion resistance)
- High mechanical stability

The pressure transmitter was verified periodically with a manometer by pushing compressed air. Re-calibration of the sensor was not required.

### 5.3.1.4    Membrane holder

The membrane holder is the place where the membrane filter is placed for the filtration test. It should avoid leakages, and not damage the membrane at all. In this research a holder for 25 mm diameter membranes was used. Several types were tested (Sterlitech - Stainless steel, Whatman GE - Poly propylene, Schleicher & Schuell - Poly propelene) with the Schleicher & Schuell membrane holder chosen.

This membrane holder was slightly modified by removing the upper inner wall of the membrane holder, so in this was the flow distribution towards the membrane is uniform and only the membrane captures all the particles in the sample water. In the set-up, the filter holder is connected with a three-way valve that connects with the syringe (sample water) and with the pressure sensor.

Nahrstedt and Camargo (2008) studied the effect of the membrane holder in SDI and MFI results. They tested three filter holders: Millipore inline 47 mm, Sartorius SM 47 mm, and Sartorius SM 25 mm. They measured up to 90 % different SDI values and 20 % MFI values for the three membrane holders when filtering the same solution. The differences were attributed to different flow distribution inside the filter holder and attributed to the effective or real filter area affected by the holder support.

### 5.3.1.5    Syringe

The water sample is placed in a disposable syringe that is attached to the piston pump. The used syringe is a BD Plastipak™ 60 ml. Syringes are cleaned by soaking with lab water before testing. A new syringe is used every 4 tests or as frequent as necessary. For "lab water" is understood water with organic matter content less than 5 µg/L and conductivity 0.055 µS/cm (or resistivity, 18.2 MΩ-cm).

### 5.3.1.6    Tubing

Tubing should be pressure resistant. The tested operational pressure values were up to 4 bar. Tubing is used to connect the three-way valve with the pressure sensor. Also, it should be resistant to chemicals, aging and abrasion. The tube length is ~20 cm with a diameter 6 mm and the brand is RAUCLAIR-E.

### 5.3.1.7    Software

The measured signal of the pressure transmitter needs to be recorded for further processing. This is done in steps:

1.  Convert milli volts to bar units
2.  Save the pressure values in a data base e.g., spreadsheet (Excel).

The pressure transmitter is connected to a computer via a modem (Endress + Hauser, FXA195 HART modem) with a USB connection. In the computer the measured voltage is transformed to pressure units by the help of a software (HART OPC server) using a calibration line. With the help of a second software (RENSEN OPC office link), the pressure values are recorded in a database for further processing.

### 5.3.1.8    Membrane cleaning and conditioning

Membrane filters must be clean and pores and surface be wet before performing the MFI-UF test. A surface that is not clean may affect the way that the fouling cake is formed on the membrane and a membrane that is not wet will required more pressure during filtration.

According to the operating instructions provided by the membrane manufacturer, the membranes (PES and RC) are coated with glycerine to prevent the membrane drying out and also sodium azide ($NaN_3$) to preserve the membrane. In order to remove the coating materials, a 24 hours soaking procedure was tested and found adequate for cleaning the coating of the membranes. In addition, the membrane resistance was measured before the MFI tests by filtering UPW through the membrane.

## 5.3.2   TESTING PROCEDURE

Independent of the membrane MWCO or the flux rate for the test, the MFI-UF testing procedure is the following:

1.  The membrane resistance is measured with UPW at the same flux as the MFI-UF test to be performed.
2.  Membrane filter is placed into the membrane holder. The active layer of the membrane is placed facing towards the water sample.
3.  Filtration flux rate is controlled manually in the pump by defining the flow rate in ml/hr. The effective membrane area must be considered when calculating the flux rate.
4.  The software for recording the pressure and time values should be started. Both, pump and data logging must start simultaneously.
5.  The fouling index ($I$) is calculated by dividing the slope of the *pressure vs. time* line over square flux and water viscosity.

$$I = \frac{Slope}{J^2 \cdot \eta}$$
Eq.   5.1

6.  Criteria to stop the test:
    a.  When cake filtration is reached (linear trend between pressure and time or the slope of fouling index and time shows no change in time),
    b.  When a minimum fouling index ($I$) value is observed;
    c.  Change in MFI value in last 5 min filtration is less than 5 % per minute.
    d.  At least 35 minutes filtration occurred.
7.  MFI-UF is calculated considering the minimum $I$ values.
8.  In order to keep MFI-UF values comparable with $MFI_{0.45}$, the MFI-UF values are standardized to reference conditions namely: viscosity at temperature of 20 °C ($\eta_{20oC}$), pressure of 2 bar ($\Delta P_0$) and surface of area of a MFI 0.45 µm micro filter ($A_0$) as shown in Eq. 5.2.

$$MFI = \frac{\eta_0 \cdot I}{2 \cdot \Delta P_0 \cdot A_0^2}$$
Eq.   5.2

In the annex 5.11.2 is shown a pressure versus time plot, fouling index versus time and MFI-UF versus time for a test with a 10 kDa PES membrane for Mediterranean sea water. The use of a pressure moving average is illustrated as well.

## 5.4   Membrane characterization

### 5.4.1   SCANNING ELECTRON MICROSCOPY

Clean membranes were randomly selected from the package (containing 10 specimens), then samples were gold coated using a sputtering coater and scanned on with a with a field emission – scanning electron microscope (Jeol JSM-7500F) at various accelerating voltages and at magnifications of up to 100,000.

Figure 5.3 shows the SEM photos for PES. Pores could only be identified down to 50 kDa. In the case of PES 100 kDa different pore diameters could be observed from 8.5 nm to 38 nm. However, a pore size distribution could not be estimated.

The RC membrane is less porous than PES membrane and has a rougher surface.

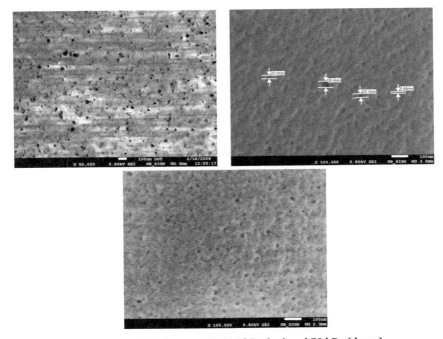

**Figure 5.3. FE-SEM pictures PES 100 kDa (up) and 50 kDa (down)**

Figure 5.4 shows the SEM pictures for RC 100 kDa membrane.

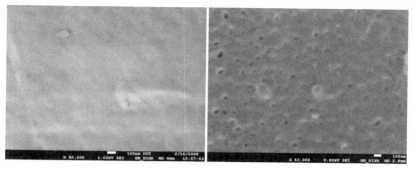

**Figure 5.4. FE-SEM picture for RC 100 kDa membrane**

Figure 5.5 shows the cross section for a PES membrane. It can be observed that this is an asymmetrical membrane with a porous support layer.

**Figure 5.5. Cross section PES membrane**

## 5.4.2   CONTACT ANGLE

The contact angle between a membrane and a droplet of an aqueous phase (e.g., water) is an indication of the overall hydrophobicity or hydrophilicity of the membrane. The lower the angle means a more hydrophilic membrane.

Hydrophobicity was estimated by the sessile drop method using a CAM 100 goniometer (KSV instruments). The goniometer with help of a video camera and software measures the left and right angle of a droplet of 2 μL of pure water on a membrane surface (Figure 5.6).

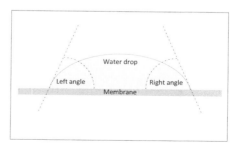

**Figure 5.6. Contact angle test as seen with the instrument (left) and scheme (right)**

Before measurement, membrane filters were soaked in ultra pure water for at least one hour, and then rinsed. Three soaking and rinsing cycles were performed to remove membrane-coating materials. Rinsed membranes were dried in a desiccator for a day and kept in petri dishes. To measure contact angle, a membrane sample piece (~1 cm$^2$) was mounted on a glass support. A 2.0 μl volume of lab water was dropped onto the membrane. Contact angle was measured within 10 seconds after the water droplet was applied (Cho, 1998, Jarusutthirak, 2002). The results of the measurements are presented in Table 5.2.

**Table 5.2. Contact angle values for PES and RC membranes**

| Material | Pore size | Left angle | Right angle | Average | Wettability |
|----------|-----------|------------|-------------|---------|-------------|
| PES | 100 kDa | 64° ± 3.9 | 65° ± 4.6 | 64° ± 4.2 | Hydrophilic + |
| RC | 10 kDa | 41° ± 0.5 | 40° ± 2.6 | 41° ± 1.5 | Hydrophilic ++ |

Both membranes materials were found to be hydrophilic; however, RC membranes are more hydrophilic than PES membranes.

In a recent study, the PES 100 kDa (56° ± 3) was reported to be more hydrophobic than RC 100 kDa (26° ± 3) (Jermann, 2008, Pieracci et al., 1999). In other studies the values for PES membranes were between 48-68° (Pieracci et al., 1999, Pontie et al., 1998, Susanto and Ulbricht, 2006). Braghetta et al. (1997) reported that the surface of a regenerated cellulose acetate UF flat disc (amicon YM series) is considered non-ionic and hydrophilic in nature and has been shown to be relatively unaffected by change in solution pH and ionic strength.

Hydrophobic membrane surfaces are often modified by blending with hydrophilic materials. The fouling potential of a hydrophobic membrane is high due to the high binding affinity of proteins and humic substances.

Adsorption of organic compounds may be related to a change in hydrophobicity / hydrophilicity of the membrane surface. Thus, the change of the contact angle may be a tool to measure adsorption. A significant increase of the contact angle for NF membranes by adsorption of natural organic matter has been reported (Roudman and DiGiano, 2000).

## 5.4.3   FOURIER TRANSFORM INFRARED SPECTROSCOPY

An ATR-FTIR Spectrum 100 instrument (Perkin Elmer) was used to measure a fourier transform infrared (FTIR) spectrum of the surface of the clean PES and RC membranes.

The system was used to determine the functional group characteristics of the membrane surface materials. Before the test, clean membranes were dried in a desiccator at room temperature for three days, and then cut into a ~1 cm$^2$ piece. The results are presented in Figure 5.7. The typical IR bands for aliphatic, aromatic functional groups, and humic substances, are summarized in Table 5.9 in the annex.

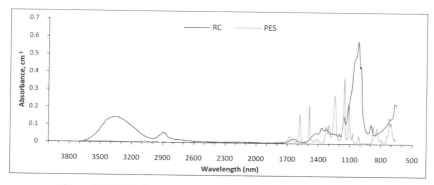

Figure 5.7. FTIR characterization of clean PES and RC membranes

The indicative peaks of RC were seen at 3400 and 1650 nm (amide carbonyl group), 2915, 1430, 1380, 1180, 1100 (aromatic double bond carbons), 1050, 1000, 930, 850, 675 (hydrocarbon, benzene ring).

The indicative peaks of PES were seen at 1300-1100 nm (ether group), 1420-1490 nm (alkanes), 1480 – 1580 nm (amide), 750 – 800 nm (ethyl group), 1325 ± 25 and 1140 ± 20 nm (sulphone group).

## 5.4.4   ZETA POTENTIAL

Zeta potential ($\zeta$) indicates the surface charge of a membrane and can be observed by measuring the streaming potential across a fluid shear plane at the surface. A streaming potential is generated when an ionic solution is forced to flow between two parallel membranes, and electrodes detect the difference in streaming potential. Zeta potential can be derived by the Helmholtz-Smoluchowski equation (Elimelech et al., 1994).

$$\frac{\Delta\phi}{\Delta P} = \frac{\varepsilon \cdot \zeta}{\mu \cdot k}$$

Eq.   5.3

Where: $\Delta\phi$ is streaming potential (mV), $\Delta P$ is forced pressure (Pa), $\varepsilon$ is the permittivity of the solution (s/m), $\mu$ is viscosity (Pa·s), and $k$ is the electrical conductivity of the solution (mS/m). The surface charge implies different fouling tendencies. Generally, membrane materials carry a negative charge or are modified to have a negative charge because NOM in water is negatively charged at neutral pH, due to phenolic and carboxylic functional groups. A negatively charged membrane, therefore, prevents rapid deposition of NOM foulants on the membrane surface by charge repellence. Studies (Childress and Elimelech, 2000, Xu and Lebrun, 1999) have determined that pH has an effect upon the charge of a membrane due to the disassociation of functional groups.

Zeta potential was measured using an SurPASS electrokinetic analyzer apparatus (Anton Paar GmbH, Austria). Membrane specimens (PES & RC,

100 & 30 kDa) were cut to fit the measurement cell and then wetted in 0.05 mM KCl solution. The zeta potentials of the membranes  were determined over a wide range of pH (2.5-12). The zeta potential was measured four times for each pH value. In all cases the correlation coefficient was more than 0.9. In all cases the acidic values were first measured and later on, with a new sample, the basic values were measured.

The results are presented in Figure 5.8. In general PES is more negatively charged than RC membranes. For the same material, different MWCO produced different zeta potential values. For PES, the 100 kDa membrane was more negative than the 30 kDa. For RC, the 30 kDa membrane was more negative than the 100 kDa. The iso-electric point for PES and RC membranes was found at acidic pH values; for RC at pH < 3, and for PES at pH < 4.5. At basic pH values > 10.5, the measured zeta potential values increased in all cases.

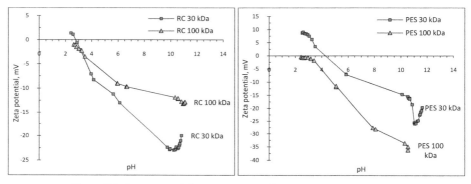

**Figure 5.8. Zeta potential measurements for RC and PES membranes**

In a recent study the zeta potential for a PES 100 kDa membrane was reported as -16.1 ± 1.0 at pH 5.4 (Jermann, 2008, Jermann et al., 2007).

Zeta potentials for most membranes have been observed in many studies to become increasingly more negative as pH is increased and functional groups deprotonate (Braghetta et al., 1997, Lee et al., 2002).

## 5.4.5   MEMBRANE RESISTANCE

The measurement of membrane resistance ($R_m$) is performed for every MFI test before measuring the sample. UPW is filtered through a membrane and the clean water pressure is obtained. Furthermore, the $R_m$ value is calculated by using the equation 5.4.

$$R_m = \frac{\Delta P}{\eta \cdot J} \qquad\qquad\qquad\qquad \text{Eq.} \quad 5.4$$

Membrane resistance values and their variations used in the experiments and collected during this research are shown in Table 5.3.

**Table 5.3. Membrane resistance values for PES and RC membranes**

| Material | MWCO, kDa | Nr. Filters | Avg. $R_m$, 1/m | Max | Min | Std Dev. |
|---|---|---|---|---|---|---|
| PES | 5 | 4 | 1.22E+13 | 1.30E+13 | 1.12E+13 | 6.2% |
| | 10 | 39 | 1.05E+12 | 1.65E+12 | 9.10E+11 | 13.8% |
| | 30 | 56 | 8.38E+11 | 7.02E+11 | 7.02E+11 | 13.8% |
| | 50 | 16 | 6.49E+11 | 7.50E+11 | 5.55E+11 | 10.5% |
| | 100 | 43 | 3.30E+11 | 4.96E+11 | 2.84E+11 | 12.9% |
| RC | 5 | 4 | 3.21E+13 | 3.47E+13 | 3.05E+13 | 7.1% |
| | 10 | 6 | 1.54E+13 | 1.64E+13 | 1.39E+13 | 6.9% |
| | 30 | 7 | 1.83E+12 | 2.06E+12 | 1.37E+12 | 12.3% |
| | 100 | 45 | 5.01E+11 | 6.47E+11 | 3.57E+11 | 16.8% |

As can be observed in Table 5.3, the smaller the MWCO the higher the membrane resistance and therefore higher pressure required (shown in Table 5.1). For RC membranes the standard deviation in the $R_m$ values ranged from 7 % for 10 kDa up to 17 % for 100 kDa. For PES membranes, where more membranes were considered in the average, the standard deviation ranged from 10.5 % for 50 kDa up to 13.8 % for 10 and 30 kDa. The average $R_m$ value for a package (10 membranes) is in general more homogeneous than the average considering several packages. This may be due to a lot-to-lot manufacturing differences while producing membranes.

Membrane resistance as expressed by Poiseuille's equation depends on thickness ($\Delta x$), tortuosity ($\tau$), porosity ($\varepsilon$) and pore size ($r_p$), as follows:

$$R_m = \frac{8 \cdot \Delta x \cdot \tau}{\varepsilon \cdot r_p^2}$$
Eq. 5.5

RC membranes showed a higher membrane resistance than PES membranes for the same MWCO. This is an indication than RC membranes most likely have a higher thickness, lower surface porosity, and/or higher tortuosity and/or smaller pore size, hence, leading to higher pressure through the membrane at same flux (shown in Table 5.1). According to Mulder (2003), a uniform molecular weight of membrane polymer does not exist but rather a molecular weight average. Hence, even though the MWCOs are the same, this does not mean that the pore size is the same as most manufacturers measure the MWCO in different ways.

Regarding to the obtained results the variations are still within the range that was reported by Cheryan in (1998a). In his study the $R_m$ values varied as much as ±25 % for the same operating conditions (temperature and pressure). Alhadidi et al. (2008) studied several 0.45 μm filters used in SDI tests and reported that there is a variation in membrane properties within a manufactured batch. The variations occurred for acrylic copolymer, cellulose nitrate, polyvinylideenfluoride, and polytetrafluoroethylene. In Alhadidi's study the variations were in pore size and roughness up to an average of 10 %

and 17 %, respectively within a batch of membranes. Less variation was observed in bulk porosity, which was lower than 5 %; variation in membranes thickness ranged from 3 to 7 %.

The variation in $R_m$ might be due to a non uniform pore size distribution and non uniform surface porosity.

### 5.4.5.1    Membrane weight

From equation 5.5, assuming that tortuosity ($\tau$), porosity ($\varepsilon$), and pore size ($r_p$) of the same batch of membrane are uniform, then the characteristics of a membrane can be indicated by the thickness ($\Delta x$) of the membrane, which can be represented by its weight.

The weights of clean membranes were measured and verified a correlation with membrane resistance (least squared $R^2$, and pearson coefficient $r$). Pearson coefficient is a measure of the correlation between two variables (e.g., X and Y). The value 0 means that there is no correlation and the closer the coefficient values to 1 indicates a strong correlation which means that the increase in membrane weight will give higher $R_m$ values in a linear relationship. In contrast, values closer to -1 indicate that the correlation is inverse, which means that the increasing membrane weight will give a proportionally lower $R_m$ value.

Table 5.4. Membrane weight of RC and PES 100 kDa

| Membrane | Avg. weight, mg | Std Dev. |
|---|---|---|
| 100 kDa RC | 66.7 | 4.6 % |
| 100 kDa PES | 86.57 | 1.2 % |

For ten RC 100 kDa membranes and ten PES 100 kDa membranes, weights were measured as reported in Table 5.4. The results revealed that the variation in membrane thickness was not substantial. PES membranes have a variation in membrane thickness lower than 2 % in comparison lower than 5 % for RC.

Furthermore, for two extra membrane sets (PES 30 kDa and RC 100 kDa), their membrane resistance and weight was measured to study a possible correlation. The results are shown in Table 5.5.

Table 5.5. Summary of $R_m$ and membrane weight of PES 30 kDa and RC 100 kDa

| | PES 30 kDa | | RC 100 kDa | |
| | Weight, mg | $R_m$, m$^{-1}$ | Weight, mg | $R_m$, m$^{-1}$ |
|---|---|---|---|---|
| Average | 86.70 | 9.46E+11 | 68.23 | 5.20E+11 |
| Max | 88.22 | 1.11E+12 | 75.1 | 5.53E+11 |
| Min | 84.5 | 8.43E+11 | 64.63 | 4.69E+11 |
| Std Dev. | 1.46% | 9.94% | 5.34% | 5.14% |
| Pearson coefficient | 0.50 | | -0.66 | |

The pearson coefficient was 0.5 for the PES 30 kDa and it was -0.66 for RC 100 kDa. The values are not distinctive enough for concluding a relation between $R_m$ and weight. In addition, the $R^2$ values were 0.25 and 0.44 for PES and RC membranes, respectively.

No correlation between membrane weight and $R_m$ was observed; this suggests that the other factors may be not uniform either (porosity, tortuosity, pore diameter).

## 5.4.6   SUMMARY

This section can be summarized as follows:

- RC and PES membranes are hydrophilic. RC is more easily wettable than PES membranes.
- PES is more negatively charged than RC membranes.
- PES membranes have lower membrane resistance values than RC membranes. This suggest that PES membranes are more porous than RC membranes for the same MWCO.
- Membrane resistances differ from batch to batch up to 15 % (Std. Dev). This is attributed to a non uniform manufacturing process.

# 5.5   Variables in the MFI-UF test

## 5.5.1   MEMBRANE PORE SIZE

The pore size range of ultrafiltration membranes is large. Pore sizes can vary from a few micrometers to nanometres. As the MFI-UF test works in a dead-end configuration, all of the particles bigger than the pore size of the membranes are retained. This means that the smaller the membrane pore size the more particles will be captured, thus creating a thicker and less porous cake. At the same time, the fouling potential of the water is proportional to the concentration of particles in the water; this means that the MFI value with a smaller pore size membrane will be higher that with a looser membrane. This is illustrated in Eq. 5.6 [(Boerlage, 2007a) and (Schippers, 2007)].

$$MFI = \frac{\eta_{200C} \cdot 90 \cdot (1 - \varepsilon) \cdot C_b}{\rho_p \cdot d_p^2 \cdot \varepsilon^3 \cdot \Delta P_0 \cdot A_0^2} \qquad \qquad \text{Eq.} \quad 5.6$$

The above formula considers the ideal case that particles are spherical.

Where:

- $\rho_p$        : Particles density forming the cake, kg/m$^3$
- $\varepsilon$        : Porosity of cake
- $d_p$        : Particles diameter, m

- $\eta_{20}$ : Viscosity at 20° C, N·s/m$^2$
- $\Delta P_0$ : Trans-membrane pressure of 2 bar as reference at 20° C
- $A_0$ : Membrane surface area of $13.8 \times 10^{-4}$, m$^2$

This trend is illustrated in Figure 4.3 where North Sea water was tested with various pore sizes.

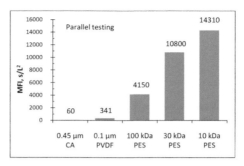

**Figure 5.9. MFI values for NSW for various MWCOs (batch 2). Measurements at constant pressure.**

It can be observed that the measured MFI value depends strongly on the pore size (MWCO) of the membrane used in the test. The feed and permeate solutions were analyzed with LC-OCD (Figure 5.10) to investigate if organic matter may contribute to the measured MFI values. Low molecular weight acids were not detected.

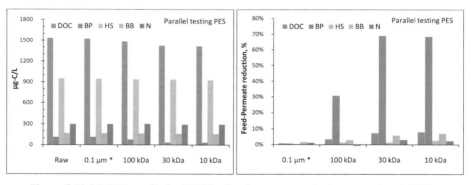

**Figure 5.10. LC-OCD results for NSW feed and permeate solutions at various MWCOs**

The DOC concentration decreased from the feed water (1.54 mg/L) by 0.7 %, 4.2 %, 7.5 %, and 7.8 % for 0.1 μm, 100 kDa, 30, kDa and 10 kDa, respectively. The organic matter fraction that was more significantly removed was the biopolymers (BP). The 100 kDa removed ~30 % of the BPs, the 30 kDa removed 69 % and the 10 kDa membrane removed ~69%. This suggests that most of the biopolymers in the feed water were bigger than 30 kDa and the rest smaller than 10 kDa. The results for the humic substances were not clear enough to suggest they contribute to the MFI values.

Comparing Figure 4.3 and Figure 5.10 right, trend similarities in biopolymers removal with the MFI values can be observed, suggesting that biopolymers contribute to the particulate fouling potential of this water. A link between biopolymers and membrane fouling has been presented in other studies (Amy, 2008, Amy and Her, 2004).

To define a membrane pore size for the MFI-UF test with the criterion "one size fits all" is incorrect as feed waters are unique. The proper membrane pore size should be selected by projecting the increase in net driving pressure and comparing with actual RO performance. Nevertheless, depending on the purpose of the measurements, some guidelines can be given. The purposes of the MFI-UF measurements can be: *i)* compare pre-treatment efficiency in removing particles or plant profiling, and *ii)* predict the rate of RO particulate fouling. For the first case, multiple MWCOs can be used to compare the efficiency of various pre-treatment processes for the removal of selected particle sizes and to determine the deposition of particles on the target membrane (Chapter 8). For the second case, the membrane used needs to be validated by the RO system operation, e.g., cleaning frequency, pressure increase or flux decline. This can be done by estimating the cleaning frequency with the help of a prediction model (Chapter 9).

## 5.5.2   FLUX RATE

In principle, the MFI-UF test can be performed at any filtration flux. However, at high flux rates two effects may play an important role in the results: flux effect on arrangement of particles during cake formation and cake compression during cake formation. An additional phenomenon might occur namely after a certain height of the cake depth filtration might occur. In this stage (final) of the filtration process small particles might be captured in the cake – similar to depth filtration in a rapid sand filter. These effects are discussed in chapter 6.

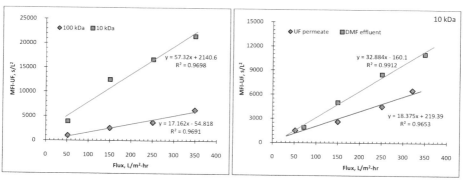

**Figure 5.11. MFI-UF values as function of filtration flux  for Mediterranean sea**

Figure 5.11 shows the MFI-UF values for Mediterranean raw seawater (left) and for UF permeate (0.02 μm pore size) and dual media filtration effluent (right) measured at various flux rates from ~50 L/m²-h up to 350 L/m²-h. A direct relation was observed between applied flux and measured MFI-UF value.

Boerlage et al. (2001a, 2004) also observed a linear trend when measuring MFI values for tap water and canal water for flux rates between 70 and 110 L/m²-h.

The applied flux may depend on the use of the measurements. In case of pre-treatment performance comparison, the MFI values can be measured at a high flux as the test duration would be short (e.g., 30 minutes) while for predicting RO operation the test can last longer as the filtration rate should be as close as possible to that of the RO (e.g., 15 L/m²-h). In annex 5.11.4 flux projections are shown for a pressure vessel containing 6 elements. The first element has the higher production while the last one can reach fluxes as low as 7 L/m²-h.

In case of ultrafiltration systems, normally operating at flux ranges between 60 – 90 L/m²-h, it was found that the high flux rate influences highly the resistance in the fouling layer. This result is described in detail in Chapter 6, and it suggests that UF systems should operate at lower flux rates e.g., < 50 L/m²-h to obtain a long term better performance than at high flux rates.

### 5.5.3   PARTICLES CONCENTRATION

Further evidence that cake filtration occurs during the MFI-UF test can be observed in the results of the MFI-UF as a function of particles concentration in the feedwater. This premise is based on the fouling index, $I$, being directly related to the concentration of particles $C_b$ (Eq. 5.7).

$$I = \alpha \cdot C_b$$                                                                Eq.    5.7

Thus, $I$ will decrease directly in proportion to an increase in the dilution factor of $C_b$ while the specific cake resistance component ($\alpha$), characteristic of a feedwater type and independent of concentration, remains constant.

In Figure 5.12, the results of the MFI-UF with dilutions of Delft canal water at an applied flux of 100 L/m²-h are shown. Linearity was found for the feedwater, with the regression coefficient calculated as 0.989.

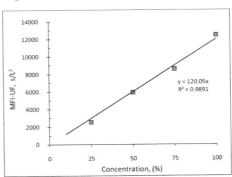

**Figure 5.12. MFI values for dilutions of Delft canal water measured with 100 kDa RC membrane at 100 L/m²-h**

Boerlage (2001a) reported a linear correlation for four different solutions in all cases between MFI-UF and concentration. Schippers and Verdouw (1980) reported, after filtering formazine solutions, that SDI is not linear with concentration while the MFI 0.45 µm is linear with concentration. A 1 mg/L of formazine had a MFI value of ~1 s/L$^2$.

## 5.5.4   MEMBRANE MATERIAL

There are several materials used in ultrafiltration such as: PES, RC, PAN, and PVDF. PVDF is produced mainly for tight MF membranes. PAN and PES are more likely for hollow fibre membranes and in various pore sizes. For this research, PES and RC membranes were tested as the range of pore size available was wider.

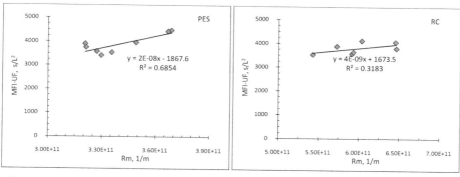

**Figure 5.13. MFI values for Delft canal water measured with 100 kDa PES (left) and 100 kDa RC (right) at 100 L/m²-h**

Figure 5.13 shows the measured MFI values for the same solution (Delft canal water) using a whole set of new membranes. For the PES membranes the average was 3,880 s/L$^2$ ± 395 (10.3 %), and for the RC membranes the average was 3,800 s/L$^2$ ± 235 (6.3 %).

Both membrane materials have an average value close to each other. RC membranes are slightly more uniform than the PES membranes when measuring MFI-UF.

## 5.5.5   OTHER EFFECTS

### 5.5.5.1   Effect of pressure on membrane material

Compaction of the membranes due to the applied pressure during filtration may occur and it may influence the MFI-UF test as, for instance, the membrane resistance in compacted membranes increases. Membrane compaction is defined as mechanical deformation of a polymeric membrane under pressure causing the porous structure to densify and consequently the flux to decline (Mulder, 2003).

To evaluate the effect of pressure on membrane compaction (increase in $R_m$), ultra pure water was filtered through PES and RC membranes (100, 30, 10 kDa). A constant pressure set-up was used for this testing. The pressure was varied between 0.5 to 3.5/4.0 bar in 0.5 bar intervals. The temperature of the feed water was maintained constant throughout the experiments ranging from 20.5–22.2° C. The flux and membrane resistance at each pressure value were measured and calculated according to eq. 5.4.

Figure 5.14 (left) shows the results of flux as function of pressure. The results indicate that 100, 30, 10 kDa PES membranes are stable over the pressure range 0.5–3.5 bar, and a linear relationship was obtained between flux and $\Delta P$ ($R^2 = 0.99$).

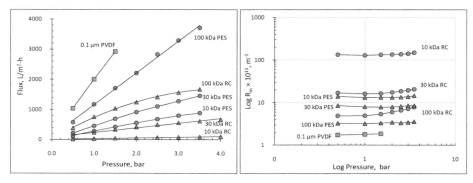

**Figure 5.14. Flux vs. Pressure (left) and Log $R_m$ vs. Log Pressure (right)**

In the case of RC membranes, for 30 and 10 kDa no significant effect of pressure on membrane compressibility was observed ($R^2 = 0.99$ linear). In contrast, the RC 100 kDa membrane showed signs of compaction as the pressure increased from 0.5–3.5 bar; the flux did not increase linearly, but started to level-off above a pressure of 1 bar. Moreover, the initial $R_m$ was increased by 38 % from 4.9 to $7.9 \times 10^{11}$ m$^{-1}$ as shown in Figure 5.14 (right).

The membrane compaction coefficient was calculated by using Eq. 5.8.

$$R_m = R_{mo} \cdot \Delta P^h \qquad\qquad\qquad \text{Eq.} \quad 5.8$$

Where: $R_m$ is the membrane resistance (m$^{-1}$), $R_{mo}$ is the membrane resistance at zero compressive pressure, $\Delta P$ is the trans-membrane pressure (bar) and $h$ is the membrane compaction coefficient.

For the 100 kDa RC membrane, a power law relationship between membrane resistance and pressure, with a *compaction coefficient* of 0.25, was observed for the range of applied pressure (0.5 and 3.5 bar). Boerlage (2001) also found a power law relationship between membrane resistance and pressure for the PAN 13 kDa. A compaction coefficient of 0.058 and 0.052 was estimated for new and used membranes, respectively. In her study, the initial membrane resistance increased by 8 % and 7 % for new and used membranes, respectively, while the applied pressure increased from 0.5 to 2 bar using RO permeate water. Boerlage concluded that this increase was not expected to have a significant effect on membrane surface properties such as pore size (Boerlage, 2001a).

### 5.5.5.2    Effect of salinity on membrane permeability

The adsorption of solutes has a negative influence on the flux because the adsorbed layer presents an extra resistance towards mass transfer and consequently contributes to a decline in flux (Mulder, 2003).

Cho et al. (2000) studied the influence of ionic strength on PEG rejection and found higher PEG rejection with higher ionic strength, thus indicating that the pore radii of the membranes are decreased by higher ionic strength. In the same study, when natural organic matter (NOM) was used, it was observed that pH and ionic strength play an important role in the charge repulsion between NOM and the membrane surface and associated NOM adsorption.

Braghetta et al. (1997) studied the permeability of a negatively charged sulfonated polysulfone NF membrane with 1 kDa MWCO and found that the permeability decreased when using ultra-pure water with different amounts of NaCl (93 – 4380 mg/L) at pH 7. The reduction of permeability was attributed to a compaction of the membrane matrix resulting from charge neutralization at the membrane surface and electric double layer compression.

The effect of salinity on the membrane was studied by measuring the MFI-UF value of synthetic seawater solution. Figure 5.15 shows two of the filtration tests with a 10 kDa membrane at 200 and 10 L/m²-h.

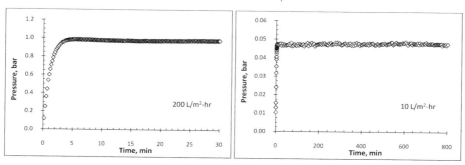

**Figure 5.15. Filtration of synthetic seawater solution through a 10 kDa membrane**

Figure 5.16 shows that the measured MFI-UF values at various NaCl concentrations were zero in all cases, thus indicating no significant effect.

### 5.5.5.3    Effect of salinity on particles

Guéguen et al. (2002) cited that increasing ionic strength is known to decrease the effective molecular size of organic molecules in solution, potentially increasing their adsorption properties on membrane sites. High ionic strength may also favour cake formation in cross flow filtration.

Typically, surface water particles are negatively charged and stable due their high zeta potential. Also, the membrane surface and pores have a negative charge and, when contacted with water, cause a polar medium which develops a double layer. Therefore, an increase of ionic strength may cause compression of the double layer around the particles and membrane surface which lead to an increase of specific cake resistance (Boerlage et al., 2003a). These considerations might be valid for hydrophobic particles which get their stability from the charge. However, hydrophilic particle, get their stability from the fact that they are surrounded and/or consists mainly of water. The stability comes from the fact that Van der Waals forces are here very weak since the attraction comes from the interaction of water molecules mainly. In coagulation "charge neutralization" does not work in this situation. Enmeshment should be strived after, which requires much higher coagulant doses.

Ribau Teixeira and Rosa (2002), reported that at high ionic strength, humic substances have a small hydrodynamic radius in solution and a large adsorbed layer thickness when adsorbed on the surface. On the other hand, at low ionic strength, humic substances have a large hydrodynamic radius and a small adsorbed layer thickness.

In Figure 5.16 the MFI-UF values for Delft canal water (diluted 4 times) and for RO feed water (North Sea, diluted from 35 to 23 g/L) are presented. The salinity of the initial solution was altered by adding concentrated solution of NaCl (99.9999 %). For both samples an increase of MFI-UF with salinity was observed.

For a salinity level and an increase similar to a 40 % SWRO recovery, the increase was about 10 % in the case of canal water and, in the case of RO feed water, the increase was about 20 %.

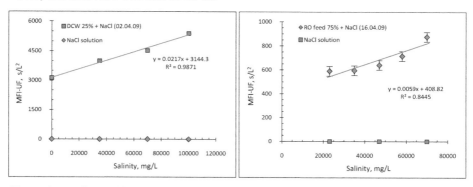

**Figure 5.16. Salinity effect on particles - MFI-UF of Delft canal water diluted to 25 %(left) and SWRO feed (right)**

From Figure 5.16 can be observed that the MFI is higher the higher the ionic strength. This suggests that salinity may play a role when comparing RO feed and RO concentrate waters when measuring the deposition factor in a real RO plant.

Boerlage et al. (2003a) tested the effect of salinity on tap water in the range 0 to 0.2 mol/L, and observed a peak value at 0.1 mol/L. Boerlage explained that ionic strength causes an initial increase in specific cake resistance due to a reduction in cake porosity which is caused by a decrease in the inter-particles distance between particles in cake filtration.

## 5.5.6   LIMIT OF DETECTION

The LOD is the concentration or amount corresponding to a measurement level (response, signal) three $s_{bl}$ units above the zero analyte (Taverniers, 2004). To measure the LOD the procedure is measure, each once, a minimum of either 10 independent sample blanks (LOD = Average + 3×StdDev) or 10 independent samples blanks fortified at lowest acceptable concentration (LOD = 3×StdDev).

Detection/results below this LOD is possible, but has a higher level of uncertainty. By using $k = 3$ times the standard deviation and a sample size (n) of at least 10, there is only a 1 % chance that a blank sample will have a higher signal than the LOD. As both $k$ and $n$ decrease, the probability that a blank sample has a higher signal that the LOD increases.

Therefore, the test in the lab is as follows. At least ten (10) blanks (lab water) were measured with various membrane MWCOs (e.g., 10, 30, 50 and 100 kDa). The mean of the signal of those blanks and the standard deviation was calculated. Results are presented in Table 5.6.

**Table 5.6. Limit of detection for MFI-UF constant flux**

| MWCO, kDa | Average, s/L$^2$ | Std. Dev. | n | LOD, s/L$^2$ |
|---|---|---|---|---|
| 100 | 13.9 | 6.8 | 10 | 34.3 |
| 100 | 22.5 | 6.5 | 27 | 42 |
| 50 | 25 | 12.5 | 10 | 62.5 |
| 30 | 35.2 | 10.5 | 10 | 66.7 |
| 10 | 15.3 | 8.7 | 30 | 41.4 |

The LOD is defined as the average of at least 10 sample blanks plus three (3) times the standard deviation. The average LOD for the n=5 batch measurements was 49.38 s/L$^2$. Therefore, the LOD was set at 50 s/L$^2$. Any MFI-UF measurement with value lower than 50 s/L$^2$ was noted as below detection limit (bdl).

# 5.6    Applications

Nevertheless, the question remains: how to translate the MFI value into reverse osmosis fouling rate? It is clear that the small particles will reach the RO units, but can they be directly translated into fouling of the RO?

For answering this question, the fouling prediction model developed by Schippers et al. (1981) was used to predict the rate of particulate fouling considering the measured MFI values and considering some other important factors: flux effects on the test and amount of particles depositing in the RO unit. This is illustrated later on 5.6.4 section.

## 5.6.1    RAW WATER COMPARISON

The MFI-UF value for seawater from various locations (Mediterranean and North Sea) were measured with a 100 kDa RC membrane at 250 $L/m^2$-h. Also, presented are the values after 0.45 μm filtration (Figure 5.17).

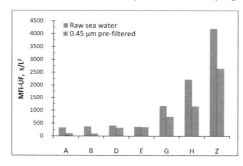

**Figure 5.17. MFI values for raw seawater from different locations. Measured with a 100 kDa RC membrane at 250 $L/m^2$-h**

The MFI values are 4 to 10 times different depending on the location and sampling period (not mentioned). These values give an indication of the particle content that the pre-treatment needs to remove before reaching a RO unit. In order to understand if these values are able to cause serious fouling, the fouling rate in RO systems can be predicted. This is illustrated in section 5.6.4 for RO systems and section 5.6.5 for ultrafiltration systems.

## 5.6.2    PARTICLE SIZE DISTRIBUTION AND FOULING

To investigate the relation between particle size and MFI-UF, North Sea water (NSW 1) was tested in series. Serial testing consisted of using the permeate water of the first filtration as a feed for the next filtration test with smaller MWCO than the previous.

Results in Figure 5.18 show an irregular trend. The MFI-UF value for particles range 0.1 μm - 100 kDa and 30-10 kDa are of the same order of magnitude and 3-4 times higher than the values for fractions 100-50 kDa and 50-30 kDa. These results illustrate a particle size distribution in the sample water. In the same way, the MFI-UF has a linear relationship with the particles concentration where MFI-UF value increases as particle concentration increases.

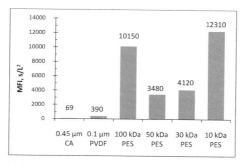

**Figure 5.18. MFI-UF results for serial fractionation of NSW 1**

A sample of each size fraction was analyzed by LC-OCD. Results are presented in Figure 5.19 left and right. The SUVA of the feed water is ~2.7 L/mg-m.

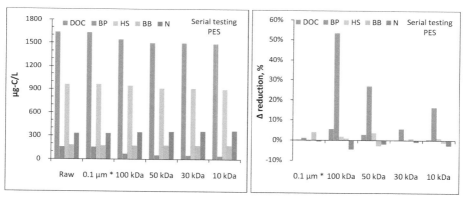

**Figure 5.19. LC-OCD results –NSW 1 – Serial Fractionation**

With respect to the feed water, a total DOC removal of ~9 % was found in the permeate of the 10 kDa membrane and the partial DOC removal was 0.2 %, 5.7 %, 8.3 %, 8.5 % and 8.7 % for 0.1 μm, 100 kDa, 50 kDa, 30 kDa and 10 kDa, respectively. The biopolymers were the OM fraction that was mainly retained by the filters (73 % in total). The partial removal of biopolymers was 1 %, 52 %, 26 %, 6 % and 16 % for 0.1 μm, 100 kDa, 50 kDa, 30 kDa and 10 kDa, respectively. Humic substances were slightly (~6 % in total) removed after the 10 kDa membrane with respect to the raw water.

A second serial fractionation with a different sample from the North Sea (NSW 2) was tested as shown in Figure 5.20. This feed water indicated that the particles retained by a MWCO of 30 kDa were the most foulant particles. Nevertheless, the water after 30 kDa membrane still has particles which produced a similar MFI-UF value as the 100 kDa membrane.

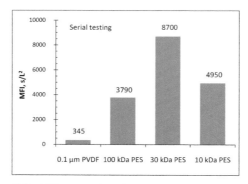

Figure 5.20. MFI-UF results of serial fractionation for NSW 2

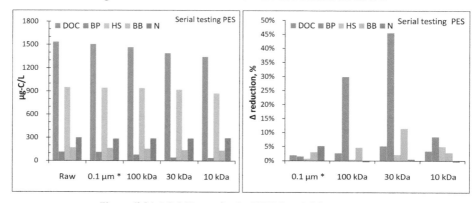

Figure 5.21. LC-OCD results for NSW 2 serial fractionation

With respect to the feed water, a total DOC removal of 13 % was found in the permeate of the 10 kDa membrane. The total removal of biopolymers after 10 kDa membrane was 66 %. Also with respect to the feed water, the total removal of biopolymers were 1 %, 31 %, 62 % and 66 % for the 0.1, 100 kDa, 30 kDa and 10 kDa, respectively. The partial biopolymers removal was 1 %, 30 %, 46 % and 8 % for 0.1 μm, 100 kDa , 30 kDa and 10 kDa respectively. In the LC-OCD test, the low molecular weight acids were not detected.

In both cases, NSW1 and NSW2, the organic matter fraction that was mainly removed by the filters was the biopolymers. For the NSW 2, there is a more clear relation between the biopolymer and humic substances removal and MFI-UF values at 100, 30 and 10 kDa; while for NSW 1, there is a high biopolymer removal and high MFI value in the 100 kDa membrane.

## 5.6.3   PLANT PROFILING

Figure 5.22 shows the MFI-UF values measured with 100, 50 and 10 kDa membranes at 250 L/m²-h along a SWRO plant treating water from the North Sea. The plant is located in The Netherlands.

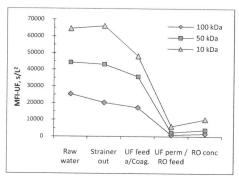

**Figure 5.22. Plant profiling with MFI-UF values measured with 100, 50 and 10 kDa at 250 L/m²-h**

The percentages in reduction of MFI values after water passing through the ultrafiltration units were 94.3 %, 93.4 % and 87.6 % for 100, 50 and 10 kDa respectively.

**Table 5.7. MFI-UF (100 kDa) values in s/L² and percentage removal**

| Date | Raw water | UF feed | UF perm | Reduction |
|------|-----------|---------|---------|-----------|
| 23.04.09 | 4310 | 2935 | 190 | 94% |
| 28.04.09 | 4840 | 4295 | 125 | 97% |
| 16.06.09 | 3800 | 3650 | 395 | 89% |
| 02.07.09 | 2950 | 2285 | 203 | 91% |
| 06.07.09 | 2840 | 2450 | 200 | 92% |
| 10.05.10 | 25340 | 17190 | 980 | 94% |

In Table 5.7 the MFI measurements, with 100 kDa membranes, are presented at various dates. Although the raw water values varied with time, the percentage decrease of MFI-UF values was, in all cases, more than 90 %.

## 5.6.4   RO PARTICULATE FOULING PREDICTION

By using the cake filtration model to predict particulate fouling on the RO elements (see Eq. 5.9) it is possible to estimate the time for a defined net driving pressure increase (15 % was considered).

$$t_r = \frac{(\Delta NDP)}{I \cdot \psi \cdot \Omega \cdot J_0^2 \cdot \eta_r} \qquad\qquad \text{Eq.   5.9}$$

Where: $\Delta NDP$ is the net driving pressure increase (bar) which typically is the criteria for cleaning in RO systems; $I$ is the fouling index $(1/m^2)$ and is calculated from the filtration test; $\Omega$ is the deposition factor (-) and is measured on-site; $\psi$ is the cake ratio factor (-).

For the projection, the following conditions were considered: temperature $= 10.5°$ C, feed pressure $= 58.5$ bar, net driving pressure $= 22.2$ bar, % increase $= 15$ % (3.32 bar), recovery $= 40$ %, flux $= 15$ L/m²-h, and deposition factor $= 1$.

**Table 5.8. Predicted cleaning frequency in RO unit for particulate fouling**

| Sample | MWCO, kDa | MFI @ 250 lmh, s/L² | MFI @ 15 lmh, s/L² | Ω | $t_r$ @ 250 lmh, months | $t_r$ @ 15 lmh, months |
|---|---|---|---|---|---|---|
| RO feed | 100 | 980 | 80 | 1 | 7.7 | 94.6 |
| | 50 | 2350 | 340 | 1 | 3.2 | 22.3 |
| | 10 | 5975 | 850 | 1 | 1.3 | 8.9 |
| Raw water | 100 | 25340 | 2816 | 1 | 0.30 | 2.7 |
| | 50 | 44285 | 4921 | 1 | 0.17 | 1.5 |
| | 10 | 64500 | 7167 | 1 | 0.12 | 1.1 |

The projected values for cleaning frequency for RO feed water and for raw water are presented in Table 5.8. For the RO feed water, in the worse case it would take ~9 months to reach a 15 % increase in NDP. Considering that raw water would be fed directly into the RO units, the predicted time for 15 % increase in NDP are 1.1 months. These values are considering that the deposition factor is 1 (all particles would accumulate on the surface of the membranes).

## 5.6.5   UF FOULING PREDICTION

Based on the measured MFI-UF values with various membranes, it is possible to project the pressure increase in the UF unit and to compare it with the real information from the plant (Figure 5.25). The projections are presented in Figure 5.23 for 1 hour filtration and in Figure 5.24 for the results with the 10 and 100 kDa membranes for various filtration times. Table 5.13 shows the projected pressure values of the calculations.

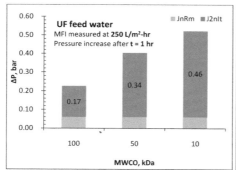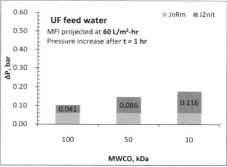

**Figure 5.23. Projected pressure increase for UF feed water after 1 hour filtration for MFI values at 250 lmh (left) and 60 lmh (right)**

The projected pressure increase with the 10 kDa membrane is higher than with 50 or 100 kDa membrane as its MFI value (48,000 s/L²) is higher than in the other two cases (35,650 s/L² and 17,190 s/L², respectively). After 1 hour filtration the pressure increase would be 0.46 bar (MFI at 250 L/m²-h) and 0.12 bar (MFI at 60 L/m²-h).

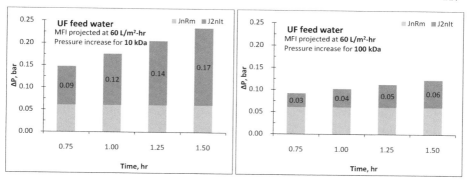

Figure 5.24. Projected pressure increase for UF feed water and MFI values at 60 lmh for 10 kDa (left) and 100 kDa (right) after various filtration times

The longer the filtration time, the higher the value of projected pressure. With 10 kDa membrane the projected pressure increase is ~3 times higher than with 100 kDa membrane.

The projected values need to be compared with data from the operator of the plant (Figure 5.25).

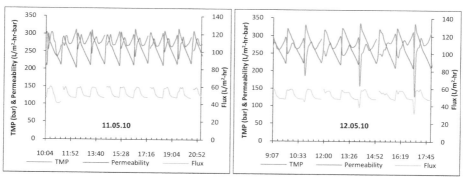

Figure 5.25. Measured trans-membrane pressure, permeability and flux for the UF system for 11.05.10 (left) and 12.05.10 (right)

It can be observed that the increase in pressure for one cycle operation is around 0.05 bar.

Comparing these values with the projected ones shows that the 100 kDa projection is in agreement (Figure 5.23 right).

## 5.7    Conclusions

- A new semi-portable set-up has been successfully developed to perform MFI-UF tests at constant flux filtration. The set-up has been used for on-site testing and for testing in laboratory.
- Two membrane materials (PES and RC) and various MWCOs (100, 50, 30 and 10 kDa) were investigated for MFI-UF tests.

- MFI-UF measured with membranes with smaller pores gives higher values. Measurements at lower flux result in lower values.
- The particulate fouling potential of Mediterranean and North Sea water was measured with MFI-UF (with 100 kDa membranes). The measured values were 2 to 8 times higher for the North Sea samples than for the Mediterranean samples.
- The biopolymers fraction in seawater organic matter was mainly removed by the membranes used in MFI-UF tests.
- Serial tests with membranes with declining pore size show that a variable particle size distribution exists.
- MFI-UF constant flux can be used for RO particulate fouling prediction.
- The limit of detection for the MFI-UF constant flux was determined for membranes of 100, 50, 30 and 10 kDa. The estimated LOD was 50 $s/L^2$.
- MFI-UF constant flux is a strong tool in prediction the development of pressure increase within a run in UF/MF systems and evaluating the efficiency of backwashing.
- An important finding was that an ultrafiltration pilot plant reduced the MFI-UF – measured with membrane with pores 100, 50 and 10 kDa – with 94 %, 93 % and 87 %, respectively.

## 5.8    Further studies

- To investigate: MFI-UF reproducibility at smaller membrane MWCOs than 100 kDa.
- To investigate whether smaller particles than 10 kDa may play an important role in RO particulate fouling.

## 5.9    Acknowledgments

Special thanks to Mr. Frederik Spenkelink for supporting in the development of the filtration set-up. To Mr. Marcelo Gutierrez for his help in programming the data acquisition software. To Mr. Frans Oostrum and Mr. Steven Mookoek (Aerospace engineering faculty, University of Delft). To Mr. Mayur Dalwani and Mr. Abdulsalam Al-hadidi (Membrane technology group, University of Twente). To Mr. Rinnert Schurer (Evides).

# 5.10  List of abbreviations and symbols

## 5.10.1  ABBREVIATIONS

| | |
|---|---|
| kDa | Kilo Dalton |
| MFI-UF | Modified fouling index – ultra filtration |
| MWCO | Molecular weight cut off |
| PES | Polyethersulfone |
| RC | Regenerated cellulose |
| RO | Reverse osmosis |
| SWRO | Seawater reverse osmosis |
| UF | Ultra filtration |
| Std. Dev. | Standard deviation |
| LOD | Limit of detection |

## 5.10.2  SYMBOLS

| | |
|---|---|
| $A$ | Effective membrane surface area ($m^2$) |
| $C_b$ | Concentration of particles in a feed water ($kg/m^3$) |
| $d_p$ | Diameter of particles forming the cake (m) |
| $I$ | Fouling index of particles in water to form a layer with hydraulic resis. ($m^{-2}$) |
| $J$ | Permeate water flux ($m^3/m^2 \cdot s$) |
| $R_m$ | Membrane resistance ($m^{-1}$) |
| $V$ | Filtrate volume ($m^3$) |
| $\alpha$ | (Average) specific cake resistance (m/kg) |
| $\varepsilon$ | Membrane surface porosity (-) |
| $\eta_T$ | Water viscosity at temperature T ($N \cdot s/m^2$) |
| $\tau$ | Tortuosity of membrane pores |

# 5.11  Annex

## 5.11.1  FTIR TYPICAL RESPONSES

Table 5.9. Typical IR spectra for aliphatic, aromatic groups, and humic substances [(Cho, 1998) in Jarusutthirak (2002).]

| Type and frequencies ($cm^{-1}$) | Assignment |
|---|---|
| Aliphatic groups | |
| 1) Hydrocarbon | |
| 2950 - 2750 | -CH, -CH$_2$ and -CH$_3$ |
| 1460 | -CH$_2$ and -CH$_3$ |
| 1380 | -CH$_3$ |
| 2) Aldehyde | |
| 2900 - 2700 | -CH stretching for aldehyde group |
| 1740 - 1730 | C=O carbonyl structure in aliphatic aldehydes |
| 975 - 780 | -CH deformation of aliphatic aldehydes |
| 3) Ketone | |
| 1715 | aliphatic ketone or carboxylic acid |
| 1250 - 1050 | C-CO-C structure in aliphatic ketones |
| 4) Amide | |

| | |
|---|---|
| 3600 - 3200 | -NH$_2$ stretching vibration of primary amide |
| 1640 | amide carbonyl group |
| 750 - 700 | -NH$_2$ /-NH structure in aliphatic primary and secondary amides |
| Aromatic groups | |
| 1) Hydrocarbon: | |
| 3050 - 3000 | -CH stretching vibration in aromatic ring |
| 1600 and/or 1500 | -C=C- stretching of aromatic ring |
| around 675 | bending and vibration of benzene ring |
| 2) Amide | |
| around 3300 | N-H stretching in secondary aromatic amides |
| 1680 - 1630 | carbonyl group of secondary amide |
| 3) Aromatic acids | |
| 3200 - 2500 | -OH stretching of hydrogen bonded carboxylic acid |
| 1690 | carbonyl group absorption of conjugated carboxylic acid |
| 1320 - 1210 | -C-O- stretching absorption for carboxylic acid |
| 4) Aromatic ketone | |
| 1700 | ketone carbonyl |
| 1600 and 1500 | C=C functional group |
| around 1715 - 1680 | carbonyl absorption of conjugated aldehyde |
| Humic substances | |
| 3400 - 3300 | O-H stretching, N-H stretching |
| 2940 - 2900 | aliphatic C-H stretching |
| 1725 - 1720 | C=O stretching of COOH and ketones |
| 1660 - 1630 | C=O stretching of amide groups (1° amide), quinone, C=O and/or C=O or H-bonded conjugated ketones |
| 1620 - 1600 | aromatic C=C |
| 1590 - 1517 | COO- symmetric stretching, N-H deformation, C=N stretching (2° amide) |
| 1460 - 1450 | aliphatic C-H |
| 1400 - 1390 | OH deformation, C-O stretching of phenolic OH, C-H deformation of CH$_2$ and CH$_3$ groups, COO- antisymmetric stretching |
| 1280 - 1200 | C-O stretching, OH deformation of COOH, C-O stretching of aryl ethers |
| 1170 – 950 | C-O stretching of polysaccharide, polysaccharide-like substances, Si-O of silicate impurities |

## 5.11.2   MFI-UF CALCULATION

**Table 5.10. MFI-UF values for various moving average for Mediterranean water (10 kDa PES, 250 L/m$^2$-h)**

| | Slope | Intercept | | | | |
|---|---|---|---|---|---|---|
| | J$^2$·h·I, bar/min | J·η·R$_m$, bar | I, 1/m$^2$ | MFI, s/L$^2$ | MIN MFI | AVG MFI |
| P vs. t | 0.006481 | 0.73436 | 2.318E+12 | 3043 | | |
| 3 min avg | 0.006081 | | 2.175E+12 | 2855 | 2855 | 3367 |
| 5 min avg | 0.006431 | | 2.300E+12 | 3020 | 3020 | 3366 |
| 10 min avg | 0.006537 | | 2.338E+12 | 3070 | 3070 | 3381 |
| 15 min avg | 0.006611 | | 2.365E+12 | 3104 | 3104 | 3402 |

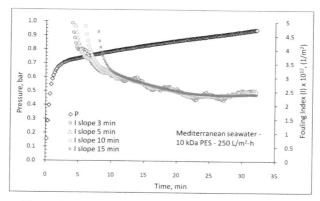

**Figure 5.26. Pressure vs. time and Fouling index vs. time**

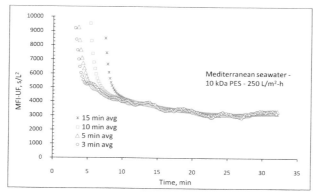

**Figure 5.27. MFI-UF vs. time**

## 5.11.3  PRESSURE TRANSMITTER

The accuracy of the pressure transmitter is presented below.

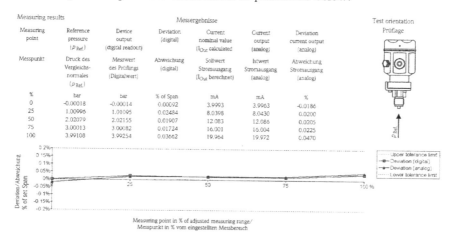

| Measuring results | | | | Messergebnisse | | | Test orientation |
|---|---|---|---|---|---|---|---|
| Measuring point | Reference pressure ($P_{Ref.}$) | Device output (digital readout) | Deviation (digital) | Current nominal value ($I_{Out}$ calculated) | Current output (analog) | Deviation current output (analog) | Prüflage |
| Messpunkt | Druck des Vergleichs-normales ($P_{Ref.}$) | Messwert des Prüflings (Digitalwert) | Abweichung (digital) | Sollwert Stromausgang ($I_{Out}$ berechnet) | Istwert Stromausgang (analog) | Abweichung Stromausgang (analog) | |
| % | bar | bar | % of Span | mA | mA | % | |
| 0 | -0.00018 | -0.00014 | 0.00092 | 3.9993 | 3.9963 | -0.0186 | |
| 25 | 1.00996 | 1.01095 | 0.02484 | 8.0398 | 8.0430 | 0.0200 | |
| 50 | 2.02079 | 2.02155 | 0.01907 | 12.083 | 12.086 | 0.0205 | |
| 75 | 3.00013 | 3.00082 | 0.01724 | 16.001 | 16.004 | 0.0225 | |
| 100 | 3.99108 | 3.99254 | 0.03662 | 19.964 | 19.972 | 0.0470 | |

**Figure 5.28. Pressure transmitter accuracy range**

## 5.11.4  SWRO DESIGN PROJECTIONS

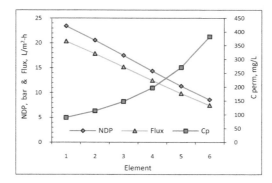

**Figure 5.29. NDP, flux and permeate concentration projections for a 15 m³/h SWRO system using SWC6 elements**

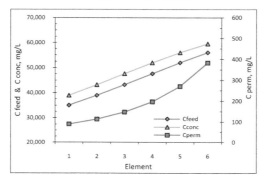

**Figure 5.30. Feed, permeate and concentrate concentration projections for a 15 m³/h SWRO system using SWC6 elements**

## 5.11.5  MEMBRANE RESISTANCE

### 5.11.5.1  Correlation of $R_m$ with filtration flux

**Table 5.11. $R_m$ of RC 100 kDa at various flux rates**

| Flux | $R_m$ 1, m$^{-1}$ (increase) | $R_m$ 2, m$^{-1}$ (decrease) | $\Delta$, % |
|------|------------------------------|------------------------------|-------------|
| 50   | 3.59E+11 | 3.49E+11 | 2.82% |
| 100  | 3.52E+11 | 3.53E+11 | 0.28% |
| 150  | 3.56E+11 | 3.60E+11 | 1.12% |
| 200  | 3.57E+11 | 3.63E+11 | 1.67% |
| 250  | 3.69E+11 | 3.71E+11 | 0.54% |
| 300  | 3.72E+11 | 3.74E+11 | 0.54% |
| 350  | 3.80E+11 | 3.80E+11 | 0% |
| Avg. | 3.64E+11 | 3.64E+11 | |
| Std Dev. | 2.81% | 3.10% | |

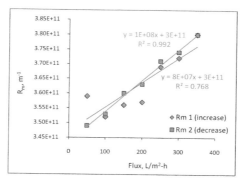

**Figure 5.31. Plot of $R_m$ values as a function of flux for RC 100 kDa (C9DN94009)**

**Table 5.12. $R_m$ of RC 100 kDa as a function of filtration flux (batch C9CN81474)**

| Flux | $R_m$ at 20° C (decrease) |
|------|------------------------|
| 500 | 3.74E+11 |
| 450 | 3.67E+11 |
| 400 | 3.62E+11 |
| 350 | 3.60E+11 |
| 300 | 3.58E+11 |
| 250 | 3.54E+11 |
| 200 | 3.52E+11 |
| 150 | 3.47E+11 |
| 100 | 3.46E+11 |
| Avg. | 3.58E+11 |
| Std Dev. | 2.6% |

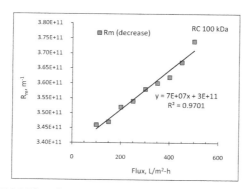

**Figure 5.32. Plot of $R_m$ values as a function of flux for RC 100 kDa**

### 5.11.5.2   Membrane resistance and weight

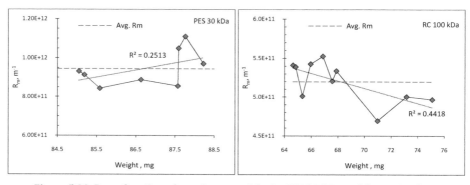

**Figure 5.33. $R_m$ as function of membrane weight for PES 30 kDa and for RC 100 kDa**

## 5.11.6   UF PROJECTIONS

**Table 5.13. Projected pressure increase for UF feed water**

| MWCO, kDa | Flux, L/m².h | MFI<br>MFI-UF, s/L² | T, °C | UF flux, L/m²-h | J$\eta$R$_m$, bar | t, hr | J²$\eta$It, bar | $\Delta$P, bar |
|---|---|---|---|---|---|---|---|---|
| 100 | 250 | 17190 | 11 | 60 | 0.061 | 1.00 | 0.17 | 0.23 |
| 50 | 250 | 35650 | 11 | 60 | 0.061 | 1.00 | 0.34 | 0.40 |
| 10 | 250 | 48000 | 11 | 60 | 0.061 | 1.00 | 0.46 | 0.52 |
| 100 | 60 | 4298 | 11 | 60 | 0.061 | 1.00 | 0.04 | 0.10 |
| 50 | 60 | 8913 | 11 | 60 | 0.061 | 1.00 | 0.09 | 0.15 |
| 10 | 60 | 12000 | 11 | 60 | 0.061 | 1.00 | 0.12 | 0.18 |
| 10 | 250 | 48000 | 11 | 60 | 0.061 | 0.75 | 0.35 | 0.41 |
| | | | | | | 1.00 | 0.46 | 0.52 |
| | | | | | | 1.25 | 0.58 | 0.64 |
| | | | | | | 1.50 | 0.69 | 0.75 |
| 10 | 60 | 12000 | 11 | 60 | 0.061 | 0.75 | 0.09 | 0.15 |
| | | | | | | 1.00 | 0.12 | 0.18 |
| | | | | | | 1.25 | 0.14 | 0.21 |
| | | | | | | 1.50 | 0.17 | 0.23 |

**NB.** MFI values at 60 L/m²-h were projected based on last year's data

# 5.12   References

ADHAM, S. & FANE, A. 2008. Cross Flow Sampler Fouling Index. California, USA: National
    Water Research Institute.

ALHADIDI, A., KEMPERMAN, A., J, C. S., WESSLING, M. & MEER, W. V. D. The influence of membrane properties on the silt density index. Membranes in Drinking water production and wastewater treatment, 2008 Toulouse, France.

AMY, G. 2008. Fundamental understanding of organic matter fouling of membranes. *Desalination*, 231, 44-51.

AMY, G. & HER, N. 2004. Size exclusion chromatography (SEC) with multiple detectors: a powerful tool in treatment process selection and performance monitoring. *Water science and technology: Water supply*, 4, 19 - 24.

BOERLAGE, S. F. E. 2001. *Scaling and Particulate Fouling in Membrane Filtration Systems*, Lisse, Swets&Zeitlinger Publishers.

BOERLAGE, S. F. E. 2007. Understanding the SDI and Modified Fouling Indices ($MFI_{0.45}$ and $MFI_{UF}$). *IDA World Congress-Maspalomas, Gran Canaria, Spain, October 21-26, 2007*, IDAWC/MP07-143.

BOERLAGE, S. F. E., KENNEDY, M., TARAWNEH, Z., FABER, R. D. & SCHIPPERS, J. C. 2004. Development of the MFI-UF in constant flux filtration. *Desalination*, 161, 103-113.

BOERLAGE, S. F. E., KENNEDY, M. D., ANIYE, M. P. & SCHIPPERS, J. C. 2003a. Applications of the MFI-UF to measure and predict particulate fouling in RO systems. *Journal of membrane science*, 220, 97-116.

BOERLAGE, S. F. E., KENNEDY, M. D., TARAWNEH, Z., ABOGREAN, E. & SCHIPPERS, J. C. 2003b. The MFI-UF as a water quality test and monitor. *Journal of Membrane science*, 211, 271-289.

BRAGHETTA, A., DIGIANO, F. A. & BALL, W. P. 1997. Nanofiltration of natural organic matter: pH and ionic strength effects. *Journal Environmental Engineering*, 123, 628-641.

CHERYAN, M. (ed.) 1998. *Ultrafiltration and Microfiltration Handbook*, USA: Technomic Publishing Company.

CHILDRESS, A. E. & ELIMELECH, M. 2000. Relating nanofiltration membrane performance to membrane charge (electrokinetic) characteristics. *Environ. Sci. Technol.*, 34, 3710-3716.

CHO, J. 1998. *Natural organic matter (NOM) rejection by, and flux decline of, nanofiltration (NF) and ultrafiltration (UF) membranes*. PhD Dissertation, University of Colorado.

CHO, J., AMY, G. & PELLEGRINO, J. 2000. Membrane filtration of natural organic matter: factors and mechanisms affecting rejection and flux decline with charged ultrafiltration (UF) membrane. *Journal of membrane science*, 164, 89-110.

ELIMELECH, M., CHEN, W. H. & WAYPA, J. J. 1994. Measuring the zeta (electrokinetic) potential of reverse osmosis membranes by a streaming potential analyzer. *Desalination*, 95, 269-286.

FLEMMING, H.-C., SCHAULE, G., GRIEBE, T., SCHMITT, J. & TAMACHKIAROWA, A. 1997. Biofouling—the Achilles heel of membrane processes. *Desalination,* 113, 215-225.

GUÉGUEN, C., BELINB, C. & DOMINIK, J. 2002. Organic colloid separation in contrasting aquatic environments with tangential flow filtration. *Water Research,* 36, 1677-1684.

JARUSUTTHIRAK, C. 2002. *Fouling and flux decline of reverse osmosis (RO), nanofiltration (NF), and ultrafiltration (UF) membranes associated with effluent organic matter (EFOM) during wastewater reclamation/reuse.* PhD Dissertation, University of Colorado.

JERMANN, D. 2008. *Membrane fouling during ultrafiltration for drinking water production - Causes, mechanisms and consequences.* PhD Dissertation, ETH Zurich.

JERMANN, D., PRONK, W., MEYLAN, S. & BOLLER, M. 2007. interplay of different NOM fouling mechanisms during ultra filtration flow drinking water production. *Water Research,* 41, 1713-1722.

KHIRANI, S., BEN AIM, R. & MANERO, M.-H. 2006. Improving the measurement of the Modified Fouling Index using nanofiltration membranes (NF-MFI). *Desalination,* 191, 1-7.

LEE, S., PARK, G., AMY, G., HONG, S.-K., MOON, S.-H., LEE, D.-H. & CHO, J. 2002. Determination of membrane pore size distribution using the fractional rejection of nonionic and charged macromolecules. *Journal of Membrane Science,* 201, 191-201.

MULDER, M. 2003. *Basic Principles of Membrane Technology,* Dordrecht / Boston / London, Kluwer Academic.

NAHRSTEDT, A. & CAMARGO SCHMALE, J. 2008. New insights into SDI and MFI measurements. *Water Science and Technology: Water Supply,* 8, 401-412.

PIERACCI, J., CRIVELLO, J. V. & BELFORT, G. 1999. Photochemical modification of 10 kDa polyethersulfone ultrafiltration membranes for reduction of biofouling. *Journal of Membrane Science,* 156, 223-240.

PONTIE, M., DURAND BOURLIER, L., LEMORDANT, D. & LAINE, J. M. 1998. Control fouling and cleaning procedures of UF membranes by a streaming potential method. *Separation and purification technology,* 14, 1-11.

RIBAU TEIXEIRA, M. & ROSA, M. J. 2002. pH adjustment for seasonal control of UF fouling by natural waters. *Desalination,* 151, 165-175.

ROUDMAN, A. R. & DIGIANO, F. A. 2000. Surface energy of experimental and commercial nanofiltration membranes: effects of wetting and natural organic matter fouling. *Journal of Membrane Science,* 175, 61-73.

SCHIPPERS, J. C. 2007. MFI more than an alternative for SDI. *International Desalination Workshop, Center for Seawater desalinatio Plants , European Desalination Society, 15-16 November 2007, Gwangiu.*

SCHIPPERS, J. C., HANEMAAYER, J. H., SMOLDERS, C. A. & KOSTENSE, A. 1981. Predicting flux decline or reverse osmosis membranes. *Desalination,* 38, 339-348.

SCHIPPERS, J. C. & VERDOUW, J. 1980. The modified fouling index, a method of determining the fouling characteristics of water. *Desalination* 32, 137-148.

SUSANTO, H. & ULBRICHT, M. 2006. Influence of ultrafiltration membrane characteristics on adsorptive fouling with dextran. *Journal of Membrane Science,* 266, 132-142.

TAVERNIERS, I., ET AL. 2004. Trends in quality in the analytical laboratory. II. Analytical method validation and quality assurance. *Trends in Analytical Chemistry,* 23, 535-552.

XU, Y. & LEBRUN, R. E. 1999. Investigation of the solute separation by charged nanofiltration membrane: effect of pH, ionic strength and solute type. *Journal of Membrane Science,* 158, 93-104.

# Chapter 6

# 6 Flux effects on cake compression in membrane filtration

Chapter 6 is based on:

SALINAS RODRÍGUEZ, S. G., KENNEDY, M. D., AMY, G. & SCHIPPERS, J. C. (2011). Flux effects on cake compression in membrane filtration. *Water Research*, submitted.

SALINAS RODRÍGUEZ, S. G., KENNEDY, M. D., AMY, G. L. & SCHIPPERS, J. C. (2011). Flux dependency of particulate/colloidal fouling in seawater reverse osmosis systems. *Desalination and Water Treatment*, in press.

SALINAS RODRÍGUEZ, S. G., KENNEDY, M. D., AMY, G. & SCHIPPERS, J. C. (2010). Flux dependency of particulate fouling in seawater reverse osmosis systems. In: EDS (ed.) *Membranes in drinking water production and waste water treatment.* Trondheim, Norway: EDS/IWA.

## 6.1   Introduction

By understanding the mechanisms of, and factors affecting, fouling in membrane filtration, a more optimum plant operation can be achieved and a more realistic assessment of fouling potential can be obtained.

This chapter deals with particulate fouling in constant flux membrane filtration. The effect of flux on the rearrangement of particles and the compression of the particles in a cake deposit is the subject of this chapter. Also, it is the purpose to link the concepts of: i) flux effect and ii) cake compression with the modified fouling index - ultrafiltration (MFI-UF), what this means for particulate fouling potential measurements, and consequences for fouling in RO and UF systems.

As early mentioned by Ruth (1935), constant flux filtration reveals at once the fact that the specific cake resistance is not a function of pressure alone but also depends to a considerable extent upon the rate of filtrate flow and velocity of solids deposition.

## 6.2   Background

### 6.2.1   PARTICULATE FOULING EQUATION IN CONSTANT FLUX FILTRATION

Flow through a reverse osmosis membrane can be described by:

$$Q_w = \frac{dV}{dt} = (\Delta P - \Delta \pi) \cdot K_w \cdot A \qquad\qquad \text{Eq.}\quad 6.1$$

where:

| | | |
|---|---|---|
| $Q_w$ | = | permeate flow (e.g., m$^3$/hr) |
| V | = | total filtered volume water (permeate) (L or m$^3$) |
| t | = | time (e.g., hour, minute, second) |
| $\Delta P$ | = | differential pressure (pressure feed - pressure permeate) |
| $\Delta \pi$ | = | difference osmotic pressure |
| | | (osmotic pressure feed – osmotic pressure permeate) |
| $K_w$ | = | permeability constant for water (m$^3$/m$^2$-s-bar) |
| A | = | surface area of the membrane(s) (m$^2$) |
| $Q_w/A$ | = | permeate flow through mem. surface area (m$^3$/m$^2$-h) |
| | = | flux (L/m$^2$-h) |
| $(\Delta P - \Delta \pi) =$ | | net driving pressure (NDP) |

In membrane technology, flux is defined as the ratio of the permeate flow and surface area of the membrane. It is expressed as:

$$J = \frac{Q_w}{A} = \frac{1}{A} \cdot \frac{dV}{dt} \qquad\qquad \text{Eq.}\quad 6.2$$

To simplify the equations we assume that $\Delta\pi$ is negligible. This assumption is valid for low salinity waters. Then,

$$J = \frac{1}{A} \cdot \frac{dV}{dt} = \Delta P \cdot K_w \qquad\qquad \text{Eq. \quad 6.3}$$

Frequently the concept of resistance $(R)$ is used, instead of permeability:

$$K_w = \frac{1}{\eta \cdot R_T} \qquad\qquad \text{Eq. \quad 6.4}$$

Where: $\eta$ is the viscosity of the water and $R_T$ is the total resistance [sum of membrane resistance $(R_m)$, pore blocking $(R_p)$ and cake formation $(R_c)$].

$$R_T = R_m + R_b + R_c \qquad\qquad \text{Eq. \quad 6.5}$$

Replacing Eq. 6.4 and Eq. 6.5 in Eq. 6.3:

$$J = \frac{1}{\eta} \cdot \frac{\Delta P}{R_m + R_b + R_c} \qquad\qquad \text{Eq. \quad 6.6}$$

When we assume that pore blocking does not play a dominant role in RO, then fouling is mainly due to cake formation. As a consequence:

$$J = \frac{1}{\eta} \cdot \frac{\Delta P}{R_m + R_c} \qquad\qquad \text{Eq. \quad 6.7}$$

Cake resistance is defined as:

$$R_c = I \cdot \frac{V}{A} \qquad\qquad \text{Eq. \quad 6.8}$$

and the fouling index $(I)$ is:

$$I = \alpha \cdot C_b \qquad\qquad \text{Eq. \quad 6.9}$$

Where: $I$ is a measure of the fouling characteristics of the water. The value of $I$ is a function of the nature of the particles and is proportional to their concentration. $C_b$ is the concentration of particles and $\alpha$ is the specific cake resistance per mg cake per m$^2$ membrane $(mg/m^2)$.

Reverse osmosis plants typically operate at constant capacity and recovery. So, the flux is constant. When membranes foul, the pressure needs to be increased, in order to keep the capacity (and flux) constant. Rewriting Eq. 6.7:

$$J = \frac{1}{\eta} \cdot \frac{\Delta P_t}{R_m + R_c} = constant \qquad\qquad \text{Eq. \quad 6.10}$$

Where: $\Delta P_t$ is the pressure at time "$t$" (which will increase). Rearranging Eq. 6.2 because flux is constant:

$$\frac{V}{A} = J \cdot t$$                                                              Eq.   6.11

and substituting Eq. 6.11 in Eq. 6.8:

$$R_c = I \cdot \frac{V}{A} = I \cdot J \cdot t$$                                         Eq.   6.12

This results in:

$$J = \frac{1}{\eta} \cdot \frac{\Delta P_t}{R_m + I \cdot J \cdot t}$$                   Eq.   6.13

Rearranging the previous equation we obtain:

$$\Delta P_t = \eta \cdot R_m \cdot J + \eta \cdot I \cdot J^2 \cdot t$$                  Eq.   6.14

Thus, $\Delta P_t$ is linearly proportional with time and is proportional with fouling index and with flux to the power two ($J^2$).

This is equation is valid for "dead end" filtration. In "cross flow" filtration only a part of the particles will deposit on the membrane surface due to the shear force of the cross flowing water.

Therefore, "$I$" has to be corrected with a deposition factor "$\Omega$". This factor is the fraction of particles which actually deposit on the membrane surface ($\Omega \le$ 1). Then, Eq. 6.14 becomes:

$$\Delta P_t = \eta \cdot R_m \cdot J + \eta \cdot \Omega \cdot I \cdot J^2 \cdot t$$     Eq.   6.15

Equations 6.14 and 6.15 do not consider that compression of the cake may occur simultaneously as the cake grows. So, in order to modify these equations to consider cake compression, the cake resistance equation (6.8 and 6.9) should be modified as it includes the only parameter affected by the cake properties. This parameter is the specific cake resistance and it was defined by Carman (1938) as expressed in Eq. 6.16.

$$\alpha_c = \frac{5 \cdot S_0^2 \cdot (1 - \varepsilon)}{g \cdot \rho_p \cdot \varepsilon^3}$$     Eq.   6.16

Where: $S_0$ is specific surface of the particles ($S_0 = 6/d_p$ for spherical particles); $\varepsilon$ is porosity; $g$ is gravitational acceleration constant; and $\rho_p$ is density of the particles.

Assuming that the cake layer consists of spherical particles with uniform density and particle diameter, the Carman-Kozeny relationship can be written as (Boerlage, 2001):

$$\alpha_c = \frac{180 \cdot (1 - \varepsilon)}{\rho_p \cdot d_p^2 \cdot \varepsilon^3}$$     Eq.   6.17

Where: $\varepsilon$ is cake porosity, $\rho_p$ is density of the particles, $d_p$ is diameter of the particles. As porosity is to the power three it plays a dominant role. The more compact a cake, the higher the specific cake resistance, and therefore the higher the cake resistance and a higher pressure is required to overcome this resistance.

## 6.2.2   CAKE DEPOSIT FORMATION

A major limiting factor in MF/UF, NF and RO is the permeate flux decline, or the feed pressure increase, with time due to fouling development on the membrane surface. Fouling is generally attributed to clay minerals, organic macromolecules, algae, bacteria, exopolymeric substances (EPS) or transparent exopolymer particles (TEP). Most fouling models that relate permeate flux to time or permeated volume are empirical and consider an exponential shape of fouling curves (Cheryan, 1998). Hermia (1982) developed a fouling model for dead-end filtration based on internal pore plugging, pore entrance blocking and cake filtration.

A common approach to predict permeate flux or feed pressure is the resistance-in-series model based on the flow of solvent through several transport layers. In this approach, the membrane is a selective barrier, where resistance $R_m$ depends upon the mechanical and chemical structure as well as on membrane thickness. Separation of a solute by the membrane gives rise to an increased solute concentration in the boundary layer at the membrane surface and an additional resistance due to concentration polarization $R_{cp}$. Adsorption and deposition of matter from the process feed within the membrane pores and on the membrane surface give rise to a fouling layer with an extra resistance $R_c$ to solvent flow. A series resistance of the membrane, boundary and fouling layers is used to relate permeate flux to the applied trans-membrane pressure (Choi et al., 2000). This was described in Eq. 6.5.

In the cake filtration model, macromolecules, particles or aggregates deposit on the membrane surface, forming a cake or fouling layer, increasing the hydraulic flow resistance due to foulant accumulation.

Several mechanisms have been described in membrane filtration, namely: depth filtration, blocking filtration (pore blocking), cake/gel filtration without compression, and cake/gel filtration with compression. With regards to depth filtration, since the membrane is very thin and this mechanism requires significant time to occur, depth filtration in a membrane is unlikely to occur. These mechanisms can be identified by plotting $t/V$ vs. $V$ in constant pressure filtration or by plotting $I$ vs. $t$ from $P$ vs. $t$ in constant flux filtration as illustrated in Figure 6.1.

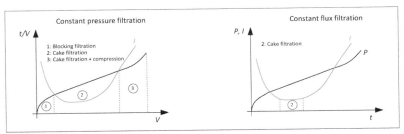

**Figure 6.1: Fouling mechanisms in constant pressure and constant flux filtration**

During the first period of filtration (region 1), blocking filtration occurs which results in a sharp increase in the slope of $t/V$ or $P$ *vs.* $t$ as can be seen in the Figure 6.1. In classic cake filtration theory, the resistance of the membrane is considered constant with time.

The modified fouling index (MFI) is based on the cake filtration mechanism. In Figure 6.1 cake filtration is considered to happen in region 2 (the middle) where the linear slope line can be observed. Additionally, in a plot of fouling index ($I$) values over time, cake filtration is observed as a minimum or stable $I$ value depending on the length of cake filtration. The MFI test assumes that at least during a period of some significance, ideal cake filtration takes place (Boerlage et al., 1998).

MFI assumes that the retention of particles is constant and specific cake resistance has a time independent permeability and uniform cake porosity throughout the entire depth of the cake, which also means that the cake is incompressible (Boerlage et al., 1998, Boerlage et al., 2002).

### 6.2.2.1   Cake filtration with compression

The current equations describing cake formation assume that the average specific cake resistance is constant over time.

According to Boerlage *et al.* (1998), very few filter cakes are incompressible since many cakes are composed of clays and microbial cells which are highly compressible. As the head loss over cake thickness is increasing, cake porosity is reduced because the particles are compressed. As a consequence, the porosity distribution becomes non-uniform across the cake layer.

Additionally, fine particles in the feed water may also reduce cake porosity even further. The fines are able to deposit in the opening inside the cake and hence block or narrow the void. Flux also may cause compression of the cake layer as the higher flux indicates that the initial void volume in the cake has decreased. This illustrates that constant flow filtration is not only a function of pressure stress but also depends to a considerable extent upon the rate of filtrate flow and velocity of solids deposition (Ruth, 1935).

As a result of decreasing porosity, the resistance of the cake layer increases due to cake compression. However, negligible compression due to flux is often assumed and the compressibility is expressed in terms of an average specific cake resistance (Boerlage et al., 1998). When the cake compressibility $\omega$ is taken into account, the empirical equation from Almy and Lewis (1912) is frequently used to relate the specific cake resistance $\alpha$ and pressure drop across the cake $\Delta P_c^\omega$, as stated in the following equation:

$$\alpha = \alpha_0 \cdot \Delta P_c^\omega$$                                          Eq.    6.18

From Eq. 6.18, there is no time variable for the cake to undergo a certain degree of compressibility and the cake is assumed to instantly be compressed (Kovalsky et al., 2008). Moreover, contrary to the assumption contained in the equation above, they showed that there was a time effect on the compressibility as can be described by the figure below.

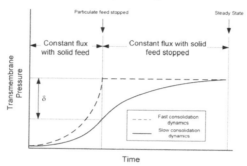

Figure 6.2. TMP profile during constant flux cake filtration followed by passage of clean solution at constant flux for arbitrary systems of fast and slow consolidation dynamics (Kovalsky et al., 2008)

As can be seen from Figure 6.2, the presence of a time lag showed that there was a time effect for a cake to reach its steady-state of compression.

### 6.2.2.2    Compressibility effect in the MFI test

Schippers (1989) introduced a correction factor to consider the compressibility of the cake in the MFI value. To measure the effect of compressibility of the cake layer in the MFI test, the behaviour of the cake layer and also the process of the cake formation and cake compression were studied further.

### 6.2.2.3    Compressible and incompressible filter cake

One of the filter cake's characteristics is its compressibility, which describes the compaction of cake structure in relation to changes in the physicochemical properties of the filtration system (Santiwong, 2008). According to Coulson and Richardson (1990) filter cakes are divided into two classes: incompressible cakes and compressible cakes, which can be distinguished by the specific cake resistance.

- For incompressible filter cakes, the specific cake resistance is not affected by the pressure differences across the cake or by the deposition of solid during filtration. The specific cake resistance is constant with time and pressure (Kovalsky et al., 2008, Rietema, 1953).
- For compressible filter cake, the specific cake resistance of the cake is affected by the pressure difference across the cake. As the pressure increases, the porosity of the compressible cake layer will decrease because of particle deformation in the cake (Carman, 1938).

Common filtration equations are based on the Darcy law:

$$\frac{dV}{A \cdot dt} = K\frac{P}{L} \qquad\qquad\qquad\qquad \text{Eq.} \quad 6.19$$

Where: $dV/dt$ is the filtration rate, $A$ is the membrane area, $K$ is the membrane or media permeability which is the inverse of resistance of the media ($R = L/K$), $P$ is the pressure drop across the media, and $L$ is the thickness of the media. The Darcy law considers that the flow is laminar and that porosity and viscosity are constant.

Ruth (1935) observed that it is practically impossible to perform constant rate filtration from the first time of filtration because there is a considerable time between the start of the filtration until the permeate flow has reached its designed constant rate. However, for constant flux filtration of an incompressible cake, the rate of filtration is considered to be constant over time by extrapolating the pressure-time backwards to the point where the permeate starts to flow. On the other hand, in the case of a compressible cake, where the porosity changes (and hence the cake resistance), the rate of filtration is not directly proportional to the pressure increase only, but also as a function of the change in the specific resistance (Coulson and Richardson, 1990).

Furthermore, specific cake resistance is not only a function of pressure stress, but also depends on the rate of filtration and velocity of solid deposition; with increasing linear velocity of solids deposition, specific cake resistance increases. As would be expected, the degree of compressibility is reduced (Ruth, 1935).

According to Carman (1938), assuming the particles are spherical in shape, the specific cake resistance ($\alpha_c$) of a cake layer as expressed in Eq. 6.16 and Eq. 6.17 depends on the particle size ($d_p$), density of the particles ($\rho_p$) and cake porosity ($\varepsilon$).

As filtration continues, especially in constant flux filtration, the specific cake resistance increases over time and leads to the pressure increase necessary to maintain the flux constant. It remains difficult to distinguish which effect is dominant between the particles size and cake porosity (Guigui et al., 2002).

Furthermore according to Coulson and Richardson (1990), almost all cakes are compressible to some extent. Even incompressible filter cake can be slightly compressed at high pressure (Carman, 1938). In addition, denser incompressible cake can be attributed to the very high flow rate or if there is vibration during cake formation. However, a small degree of compressibility might be ignored for purposes of approximation.

### 6.2.2.4   Reversible and irreversible filter cake compression

Another filter cake classification is based on its elasticity. For elastic cake, the specific cake resistance varies with the applied pressure and will be reversible. On the contrary, for inelastic cake, the specific cake resistance varies with pressure but is irreversible (Carman, 1938). The specific cake resistance will be determined by the highest pressure drop over the cake layer and will not return to its lower resistance even though the pressure subjected to the cake has been eliminated or decreased. If the reversibility of an elastic cake layer is attributable to the compression of the particle or material itself (Coulson and Richardson, 1990), then the irreversibility of an inelastic cake is caused by the breakdown of the original particle or packing of the cake during compression (Carman, 1938).

Reversibility of compression of a cake layer from coagulated water was observed by Guigui *et al.* (2002). In the filtration of distilled water over pre-deposited floc cake layer, the resistance was stable over time. Furthermore, if the pressure over the compressible cake layer is increased, the specific cake resistance is also increased. When the pressure is reduced or removed, the cake is relaxed and regains a more porous structure. As a result, the specific cake resistance of cake layer returns to its initial value and thus the compression of cake layer is reversible.

### 6.2.2.5   Flux effect and compression during cake formation

In order to study cake formation in membrane filtration it was hypothesized that the specific cake resistance is influenced by:

- flux rate at which filtration occurs at which the cake is formed.
- compression of the cake deposit.
- both occur simultaneously

The flux effect and the compression effect are illustrated in Figure 6.3.

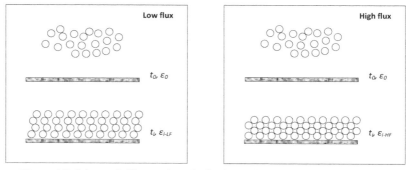

**Figure 6.3. Schematic illustration of cake formation at low flux and high flux**

The flux rate affects directly the internal arrangement of the particles in the cake. This internal rearrangement may occur simultanelously with cake compression. As can be observed in Figure 6.3, considering that the same volume of water was filtered, the cake porosity in the cake formed at high flux is lower than the porosity of the cake formed at low flux, this is $\varepsilon_{i\text{-}HF} < \varepsilon_{i\text{-}LF}$. or expressed in terms of specific cake resistance, $\alpha_{i\text{-}HF} > \alpha_{i\text{-}LF}$.

# 6.3   Materials and methods

## 6.3.1   CONSTANT FLUX FILTRATION SET-UP

This set-up has been described in detail in chapter 5. A scheme of it is presented in Figure 6.4

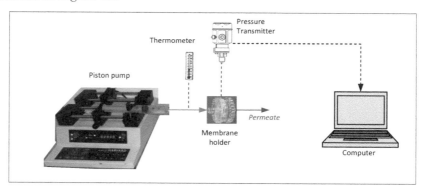

**Figure 6.4. Constant flux filtration set-up**

## 6.3.2   MEMBRANE RESISTANCE

Membrane resistance was measured with ultra-pure water (UPW). Three different MWCOs (100, 50 and 10 kDa) were used for the tests with seawater, and a 100 kDa membrane was used with Delft canal water.

UPW was produced in a multi stage process: Delft tap water was passed through a RO system (Rossmark), then filtered through GAC and ion exchange, and finally through a second RO system (Rossmark).

The membrane resistance was calculated with the help of the following equation:

$$R_m = \frac{P}{\eta \cdot J}$$                                                   Eq.   6.20

As the tested water was free of foulants, the pressure and flux were constants.

## 6.3.3   CAKE RESISTANCE

Raw water from the North Sea (TDS ~35 g/L, DOC ~1 mg/L) and Delft canal water (TDS ~0.7 g/L, DOC ~16 mg/L) were used in the compressibility tests.

In all cases, the same volume of sample water (seawater or canal water) was filtered at constant flux to build up a cake deposit on the surface of the membrane at 20 and 200 L/m$^2$-h.

After every filtration test, the cake resistance was calculated from the resistance in series model equation considering that pore blocking is negligible as illustrated in Eq. 6.21.

$$R_c = \frac{P_t}{\eta \cdot J} - R_m$$                                            Eq.   6.21

The fouling index ($I$) was calculated from the slope of the linear region of the $P$ vs. $t$ curve. From Eq. 6.14, the slope is equal to the product $\eta \cdot I \cdot J^2$. A conversion factor between MFI and $I$ can be calculated from the reference conditions of membrane area, pressure and temperature. $I = 3.8 \times 10^8 \times MFI$ (m$^{-2}$).

## 6.3.4   FLUX EFFECT AND CAKE COMPRESSION RESISTANCE

In all cases, the same volume of sample was filtered.

Compression effect:

- A cake was formed (with seawater or Delft canal water) at low flux (e.g., 20 L/m$^2$-h). This is, $R_{20\ Seawater/DCW}$.
- Immediately after, a synthetic solution (SS) was filtered through the formed cake at a higher flux (e.g., 200 L/m$^2$-h) than the formation of the cake. This is, $R_{200\ SS}$.
- The increase in resistance was considered due to pure compression of the cake ($R_{Compression}$).

- This is, $R_{Compression} = R_{200\ SS} - R_{20\ Seawater/DCW}$, resistance due to compression of the cake at flux 200 L/m$^2$-h on cake formed at 20 L/m$^2$-h.

Flux effect:

- A cake was formed (with the same seawater or Delft canal water) at high flux (e.g., 200 L/m$^2$-h).
- Resistance due to flux effect ($R_{Flux\ effect}$) was considered the difference of the resistance at high flux minus the resistance after filtering a synthetic solution earlier.
- This is, $R_{Flux\ effect} = R_{200\ Seawater/DCW} - R_{200\ SS}$.

The synthetic solution was prepared to have the same *ionic strength* (TDS ~35 g/L for seawater and TDS ~0.7 g/L for canal water), and same *ion content* as in the respective sample water. It was prepared free of organic matter and verified that it had no fouling potential; this was verified by filtering the solution and monitoring the pressure (see annex 6.8.2) vs. time with a 10 kDa membrane.

Immediately after building up the cake deposit at flux of 20 L/m$^2$-h with the sample water on the surface of the membrane, the synthetic solution was filtered at various flux rates (20, 80, 140, 200 L/m$^2$-h).

Figure 6.5 illustrates the followed procedure.

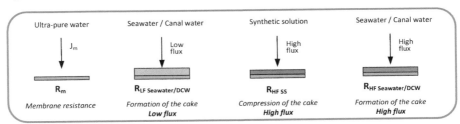

**Figure 6.5. Scheme of the flux effect and compressibility tests at constant flux**

It is important to remark that in all filtration tests, the *same volume of water was filtered* to build up the cake deposit on the membrane at *constant flux*.

## 6.3.5   ON SITE TESTING LOCATIONS

Three different locations were selected for the tests. Two are placed along the Mediterranean Sea and one is located on the North Sea. The water samples were taken from the intakes of each desalination plant. The locations and pre-treatments are briefly described on Table 6.1.

**Table 6.1. Description of the tested locations**

| Location | Intake | Pre-treatment | Comment |
|---|---|---|---|
| A (Northern Mediterranean water) | Submerged pipe (next to the shore) | Strainer – UF (0.01μm) | Flux ~57 L/m$^2$.h |

| A (Northern Mediterranean water) | Submerged pipe (next to the shore) | Strainer – Coagulation + Dual media filter | DMF filtration rate ~7.9 m³/m²-h. 2 mgFe³⁺/L + 0.2 mg Polymer/L DMF = Anthracite and Sand |
|---|---|---|---|
| B (North-Western Mediterranean water) | Submerged pump (L = 2.5 km) | UF (0.02 μm) | Flux ~50 L/m².h |
| C (North Sea water) | Submerged pipe (L = 100 m) | Strainer – UF (~300 kDa) | Flux ~60 L/m².h |

A summary of the water properties is presented in Table 6.2. Location C has the highest DOC concentration, while the location B has the lowest DOC concentration.

**Table 6.2. Summary of water characteristics**

| Sample | Location | DOC (mg/L) | pH | T (°C) | SUVA (L/mg-m) | EC (mS/cm) | Turbidity (NTU) |
|---|---|---|---|---|---|---|---|
| Raw water | A | 1.2 | 8.21 | 18 | 0.8 | 57.1 | |
| UF permeate | A | 0.85 | 8.2 | 18 | 0.7 | 57.1 | |
| Coag+DMF effluent | A | 0.77 | 7.8-8.0 | 18 | 0.6 | 57.1 | |
| Raw water | B | 0.75 | 8.1 | 16.4 | 0.55 | 57-58 | 0.5-1.5 |
| UF permeate | B | 0.72 | 8.0 | 16.4 | 0.45 | 57-58 | |
| Raw water | C | 1.45 | 8.1 | 13.5 | 2.3 | 48.5 | 8-12 |
| UF permeate | C | 1.3 | 6.5 | 13.5 | 2.15 | 48.5 | |

To reduce the impact of variations in water quality, as many tests as possible were performed on the same day.

## 6.4    Goals and objectives

This study has the following objectives and goals:

### 6.4.1    GOALS

- To study the effect of flux and cake compression in membrane filtration and their implications in measuring the "real" particulate fouling potential of seawater.

### 6.4.2    OBJECTIVES

- To measure the cake resistance at various flux rates for seawater.
- To measure the effect of flux rate on cake formation.
- To measure cake compression in constant flux filtration.
- To measure cake compression over time.
- To link the measured effects to: particulate fouling indices, particulate fouling in SWRO systems and in UF systems.

## 6.5  Results and discussion

In this section the several water samples were tested at various fluxes with different membrane MWCOs.

### 6.5.1  CAKE COMPRESSIBILITY STUDY

#### 6.5.1.1  Cake resistance as a function of flux

The cake resistance is defined by Eq. 6.22. It is a function of the fouling index of the water ($I$), the filtration flux ($J$) and the filtration time ($t$).

$$R_c = I \cdot \frac{V}{A} = I \cdot J \cdot t \qquad\qquad\qquad \text{Eq.} \quad 6.22$$

$I$ is defined as:

$$I = \alpha \cdot C_b \qquad\qquad\qquad \text{Eq.} \quad 6.23$$

Assuming that no effect of flux and no compression, in the case of filtering the same solution at various flux rates or various filtration times, the fouling index can be taken as constant ($C_b$ is the same). In the case of filtering the same solution for a fixed constant volume at various flux rates, the filtration time at high flux is short and the filtration time at low flux is long. Considering this, the filtration of a constant volume of the same solution at various flux rates would produce no change in the cake resistance (Figure 6.6 left).

The resistance of cake deposit (seawater) was measured for the same filtration volume at various constant flux rates ranging from 10 - 200 L/m²-h. The obtained results are presented in Figure 6.6 (right).

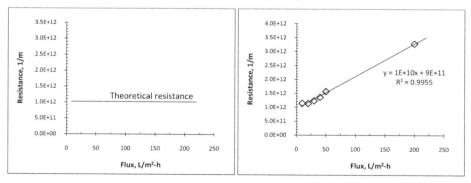

Figure 6.6. Theoretical (left) and measured (right) cake resistance versus filtration flux

From the measurements, it can be seen that there is a direct relation between filtration flux and cake resistance ($R_c$). For instance, $R_c$ increased by 1.2 times from flux 10 to 20 L/m²-h and $R_c$ increased by 5 times from flux 20 to 200 L/m²-h.

These results pose the question as to why this increase in resistance is occurring if the same volume of water was filtered? The specific cake resistance $(\alpha_c)$ is increasing due to a decrease in porosity $(\varepsilon)$ with increasing flux (see Eq.6.17). Is this decrease in porosity due to pure compression, due to the effect of flux rate on cake formation (denser rearrangement), or both?

From the measurements, it can be observed that there was cake compression/flux effect even at very low flux; and this suggests that cake compression/flux effect occurs simultaneously with cake formation. Thus, filtration flux affects the cake formation resistance.

In the following sections the flux effect on cake resistance and the compression effect on the cake resistance are presented for seawater and for canal water.

## 6.5.2    CAKE COMPRESSIBILITY AT CONSTANT PRESSURE

It was hypothesized that the cake deposit could be compressed due to the pressure during the cake filtration phase. Thus, to assess the cake compressibility factor for North Sea water (NSW), a 50 kDa PES membrane was tested at different pressures ranging from 0.5 to 3 bar. Figure 6.7 shows the results of fouling index $(I)$ values at different pressures $(P)$.

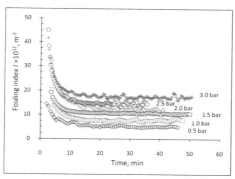

**Figure 6.7. Fouling index ($I$) vs. time for NSW at various pressure values**

The initially high $I$ values in Figure 6.7 might be related to pore blocking, or an artefact due to initial lower pressure at start. Furthermore, the $I$ values were calculated using $t/V$ versus $V$ and not $dt/dV$ versus $V$.

The fouling index $(I)$ is higher the higher the feed pressure increased (see Table 6.3), suggesting that cake compression occurred as the pressure increased from 0.5 to 3.0 bar.

**Table 6.3. Fouling index (*I*) vs. pressure for NSW**

| $\Delta P$ (bar) | $I \times 10^{12}$, (m$^{-2}$) | % increase |
|---|---|---|
| 0.5 | 5.2 | |
| 1.0 | 7.6 | 48% |
| 1.5 | 10.46 | 103% |
| 2.0 | 12.4 | 140% |
| 2.5 | 14.5 | 182% |
| 3.0 | 17.1 | 231% |

Figure 6.8 shows a log-log plot of $I$ *vs.* $P$ for NSW, which was used to calculate the compressibility coefficient ($\omega$) for the NSW water. Boerlage (2001) claimed that a high applied pressure may cause the particles and colloids in low salinity water to compress, forming a denser cake layer due to a reduction in cake porosity. Consequently, the specific cake resistance ($\alpha$) and fouling index ($I$) may increase. Also, a high flux caused by high applied pressure most likely causes cake compression, hence, the cake compression could be attributed to either increase applied pressure and/or flux increase. The average calculated $\omega$ factor for NSW was 0.67.

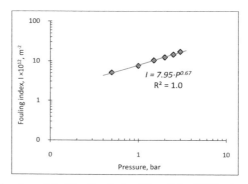

**Figure 6.8. Log-log plot of fouling index (I) vs. P for NSW using 50 kDa PES**

Figure 6.9 presents a comparison of the two approaches: (i) assuming incompressible cake ($\omega = 0$) and (ii) incorporating the compression factor ($\omega = 0.67$) for NSW in calculating MFI-UF value according to $I = \alpha \cdot C_b \cdot \Delta P_c^{\omega}$ which, at a reference condition of 2 bar, is equal to 16,195 s/L$^2$. This indicates that when compression effects are ignored, the MFI-UF value is underestimated by 60% at 0.5 bar and overestimated by 36 % at 3 bar. Hence, it is necessary to incorporate the compressibility coefficient in the MFI-UF calculation for NSW. On other hand, the compressibility factor mainly depends on the type and quality of feed water and could be different from sea to other sea in the world.

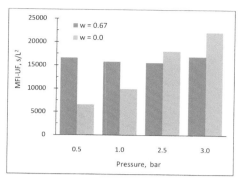

**Figure 6.9. MFI-UF vs. pressure for NSW for 50 kDa PES with/without compression factor**

Boerlage (2001) found the compressibility factor of tap water to be 0.82 and she attributed this to the large amount of hydrated colloids in treated surface water in the Netherlands. In her study, the compressibility factor after slow sand filtration was 0.68 which is very close to the compressibility factor of NSW. A compressibility factor of 0.75 is assumed to be a global compressibility coefficient after estimating compressibility for different surface feed waters. A higher compression factor indicates higher cake compression.

Tiller and Cooper (1960) studied this phenomenon in industrial filtration and described it in terms of porosity in different layers of the formed cake. After starting filtration, at each instant of time, the porosity drops throughout the cake until the membrane is reached where it has its least value. The cake thickness increases, and at a given distance from the cake surface, the porosity increases as time progresses. As the pressure drop across the cake increases, the porosity at the membrane surface decreases and eventually reaches a minimum value equal to a porosity determined by the maximum applied pressure (Tiller and Cooper, 1960).

Theoretically in pressure-driven membrane filtration, particles and colloids of feed water are transported to the membrane surface by the permeate flow. Consequently, as particles continue accumulating on the membrane surface, a cake layer forms, causing an increase in hydraulic resistance of the cake layer resulting in a decrease of permeate flux. The pressure drops across the membrane and cake can be expressed by the following equation:

$$\Delta P_T = \Delta P_m + \Delta P_c$$                                                                Eq.    6.24

Where: $\Delta P_T$ is the total applied pressure drop (bar or $N \cdot m^{-2}$), $\Delta P_m$ is the pressure drop across the membrane (bar or $N \cdot m^{-2}$) and $\Delta P_c$ is the pressure drop across the cake layer (bar or $N \cdot m^{-2}$).

In the case of a UF/MF membrane, at high applied pressure the cake layer resistance $(R_c)$ will dominate the resistance across the membrane and cake layer as illustrated in Figure 6.10 where the same volume of water was filtered in each case. Then, the membrane resistance $(R_m)$ could be negligible compared to the cake layer.

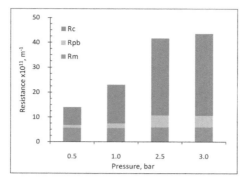

Figure 6.10. Resistance vs. pressure for North Sea water - 50 kDa PES

The results in Figure 6.10 and Table 6.4 show that membrane resistance at low applied pressure is not negligible where $R_m$ is 40 % of the $R_T$. On other hand, at 3 bar, $R_m$ could be assumed negligible where $R_m$ is only 13.6 % of the $R_T$. This assumption could be applicable for high MWCO but not for lower MWCO where $R_m$ is very high.

Table 6.4. Summary results of Resistance for NSW

| $\Delta P$, bar | $R_m \times 10^{11}$, $m^{-1}$ | $R_b \times 10^{11}$, $m^{-1}$ | $R_c \times 10^{11}$, $m^{-1}$ | $R_T \times 10^{11}$, $m^{-1}$ | $R_m / R_T$ % |
|---|---|---|---|---|---|
| 0.5 | 5.65 | 1.10 | 7.25 | 14.00 | 40 % |
| 1.0 | 5.55 | 1.89 | 15.56 | 23.00 | 24 % |
| 2.5 | 5.85 | 4.87 | 30.91 | 41.63 | 14 % |
| 3.0 | 5.92 | 4.73 | 32.84 | 43.49 | 13.6 % |

## 6.5.3   SEAWATER

### 6.5.3.1   Effect of filtration flux on cake resistance

A schematic of the tests is presented in Figure 6.11. First a fouling layer (cake deposit) was created on the surface of the membrane; for this the seawater sample was filtered at low flux (20 $L/m^2$-h). Immediately after, a synthetic solution was filtered at various constant flux rates (20, 80, 140 and 200 $L/m^2$-h). In this way, the extra resistance due to filtration of synthetic water was measured in a previously formed cake deposit.

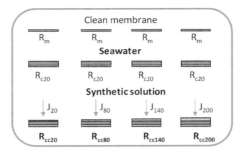

**Figure 6.11. Schematic of the test to measure flux effect**

The results of the tests with a 50 kDa PES membrane are presented in Figure 6.12 according to cake formation flux over synthetic solution filtration flux, e.g., 20/20.

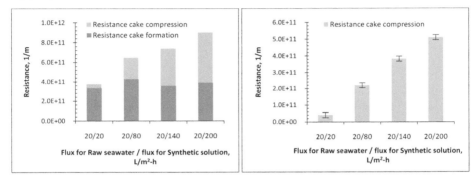

**Figure 6.12. Resistance due to cake formation and resistance due to flux effect (50 kDa PES)**

The average cake resistance was $3.78 \times 10^{11}$ m$^{-1}$ ± $4.05 \times 10^{10}$ (± 10.7 %). The deviation in the cake resistance might be due to non uniform membrane properties (surface porosity, membrane thickness, tortuosity). The membrane resistance values are not plotted in the figure [$R_m = 6.92 \times 10^{11}$ m$^{-1}$ ± $7.8 \times 10^{10}$ (± 11.4 %)].

The filtration of synthetic solution on the previously formed cake deposits illustrates the effect of pressure (due to increased flux) on cake resistance. This effect can be attributed to compression of the cake layer after filtering the synthetic solution. A denser layer of particles indicates a less porous cake and an increase in specific cake resistance.

The resistance due to filtration of synthetic solution (flux effect) increased directly with the applied flux from 20 up to 200 L/m$^2$-h. With respect to the values for 20 L/m$^2$-h, the increase was 5.4, 9.3 and 12.5 times for 80, 140 and 200 L/m$^2$-h, respectively.

The cake resistance increased even at low filtration flux, 11 % at 20/20. The increase was 34 %, 52 %, and 57 % for 20/80, 20/140, and 20/200, respectively.

The importance of using a synthetic solution for the "flux effect" tests is illustrated in annex 6.8.1. When using ultra pure water (UPW), instead of increasing the cake resistance the opposite effect was observed; this might be related to the big difference in ionic strength between the formed cake and the solution used for filtration.

### 6.5.3.2    Filtration flux effect and compression effect on cake resistance

In this case, we compared the cake resistance in a (almost) non-compressed cake (formation at low flux) with the resistance in a compressed one (formation at high flux), and filtering synthetic solution through these cakes.

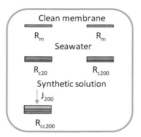

**Figure 6.13. Schematic of the test to measure compression effect**

The obtained results are presented in Figure 6.14. The cake deposits were formed at 20 and 200 L/m²-h. The cake resistance at 200 L/m²-h was 3.6 times higher than at 20 L/m²-h.

In the cake formed at low flux (20 L/m²-h, ignoring compression effect), the increase in resistance after filtering synthetic solution at 200 L/m²-h was 57 % with a total resistance = $9.01 \times 10^{11}$ m⁻¹ (cake resistance + flux effect). The resistance in the cake formed at 200 L/m²-h was $1.41 \times 10^{12}$ m⁻¹ (cake resistance + flux effect + compression effect). Comparing these two values provides an indirect measurement of the compression effect on cake formation; the difference is $5.07 \times 10^{11}$ m⁻¹ (50 %).

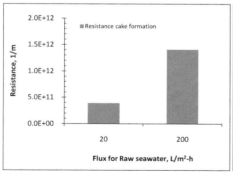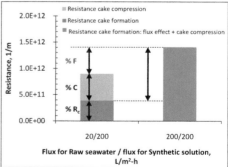

**Figure 6.14. Cake formation resistance (Left) and resistance due to flux effect and compression effect (right)**

The previous results suggest that the cake formed at 200 L/m²-h has its resistance due to 28 % "particles", 36 % flux effect on particles arrangement, and 36 % compression of cake deposit.

In the same manner as above, similar testes were followed for intermediate flux rates (80 and 140 L/m²-h). The cake deposit built up at low flux (20 L/m²-h) was taken as reference, assuming that cake compression was minimal. The filtration flux and compression effects were measured and identified for various cakes formed at different constant flux rates (80, 140 and 200 L/m²-h).

The results are presented in Figure 6.15. The cake resistance increased with the filtration flux. The flux effect also increased directly with the filtration flux.

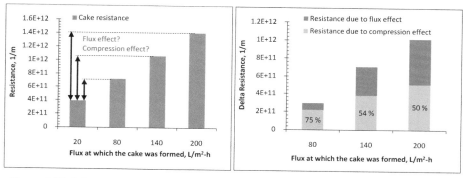

**Figure 6.15. Cake resistances (left) and resistance due to compression effect and resistance due to flux effect (right)**

As the flux for building up the cake deposit increases, both effects increase in magnitude, but in percentage the compression effect decreases (75 %, 54 %, and 50 % for 80, 140, and 200 L/m²-h, respectively) and the flux effect increases (25 %, 46 %, and 50 % for 80, 140, and 200 L/m²-h, respectively).

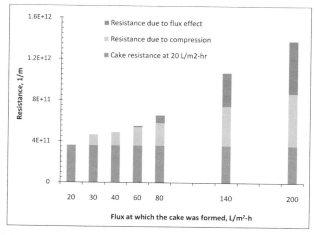

**Figure 6.16. Resistance as function of flux for flux effect and compression effect**

These results suggest that at flux rates of less than 60 L/m²-h the flux effect in the seawater is not significant as illustrated in Figure 6.16.

### 6.5.3.3    Cake compressibility as a function of MWCO

In the previous section the tests were performed with a 50 kDa membrane. However, the membrane pore size or MWCO plays a role in the cake formation as the smaller the MWCO the more particles will be captured and therefore a thicker and perhaps a more compact cake can be expected. To study the effect of the MWCO the same procedure as before was followed with a 10 and 100 kDa membrane.

In all cases the same volume of water was filtered. The results are presented in Figure 6.17 and Table 6.5.

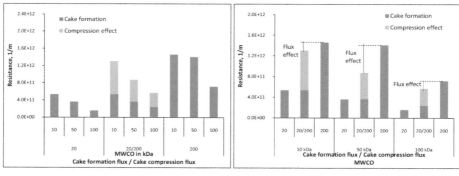

**Figure 6.17. Filtration resistances for cake deposit and compressed cake deposit as function of flux (left) and MWCO (right)**

It can be observed that when the cake was formed at 20 L/m²-h, it is clear that the cake resistance is higher for 10 kDa than for 100 kDa (2.8 times) and than for 50 kDa (1.5 times). At high flux (200 L/m²-h) the flux effect is more significant and higher for the 10 kDa membrane than for the 100 kDa membrane.

By comparing the 20 and 20/200 tests it is possible to measure the effect of compression on the cake deposit. By comparing the 20/200 test with the 200/200 test it is possible to observe the effect of flux on particle arrangement on cake resistance. Regarding the flux effect, for the 10 kDa membrane the difference is ~12 %. For the 100 kDa membrane the difference was 26 % and for 50 kDa membrane the difference was 60 %.

**Table 6.5. Resistances for clean membrane, cake deposit and flux effect - Seawater**

| MWCO, kDa | Formation / Compression flux rates | Rm, 1/m | Rcf, 1/m | R comp., 1/m | R flux, 1/m | Ratios id | Ratio | Cake formation | Compr. effect |
|---|---|---|---|---|---|---|---|---|---|
| 10 | 20/20 | 9.39E+11 | 5.35E+11 | 5.97E+10 | | a | | | |
| | 20/200 | 9.77E+11 | 5.35E+11 | 7.70E+11 | 1.55E+11 | b | b/a | 1.0 | 12.9 |
| | 200/200 | 9.27E+11 | 1.46E+12 | 4.48E+11 | | c | c/b | 2.7 | |
| 50 | 20/20 | 6.40E+11 | 3.63E+11 | 4.10E+10 | | d | | | |

| | 20/200 | 6.22E+11 | 3.63E+11 | 5.10E+11 | 5.35E+11 | e | e/d | 1.0 | 12.5 |
| | 200/200 | 8.16E+11 | 1.41E+12 | 4.34E+11 | | f | f/e | 3.9 | |
| 100 | 20/20 | 3.15E+11 | 1.58E+11 | 2.76E+10 | | g | | | |
| | 20/200 | 3.61E+11 | 2.37E+11 | 3.33E+11 | 1.47E+11 | h | h/g | 1.5 | 12.1 |
| | 200/200 | 3.72E+11 | 7.17E+11 | 3.18E+11 | | i | i/h | 3.0 | |

From the previous results, it is suggested that MWCO plays a role in cake formation as not only is the cake resistance higher at smaller MWCOs but also the flux effect has a higher impact on the cake formation. The compression effect seems to be higher at low MWCO and the flux effect on particle arrangement showed no clear trend regarding the role of MWCO.

## 6.5.4 CANAL WATER

Delft canal water (DCW) was used to study the flux effect and compressibility effect on cake formation as well.

A 100 kDa membrane was used during the tests. The average membrane resistance was $5.0 \times 10^{11}$ m$^{-1}$ $\pm 1.1 \times 10^{11}$ ($\pm 22$ %).

The obtained results are presented in Table 6.6 and Figure 6.18.

**Table 6.6. Resistances for clean membrane, cake deposit and compression effect - Canal water**

| MWCO, kDa | Formation / Compression flux rates | Rm, 1/m | Rcf, 1/m | Rcc, 1/m | Ratios id | Ratio | Cake formation | Compr. effect |
|---|---|---|---|---|---|---|---|---|
| 100 | 10/20 | 6.47E+11 | 4.94E+11 | 1.04E+11 | a | a/b | 0.9 | 3.3 |
| | 20/20 | 5.56E+11 | 5.79E+11 | 3.14E+10 | b | | | |
| | 20/200 | 3.99E+11 | 6.82E+11 | 9.07E+11 | c | c/b | 1.2 | 28.9 |
| | 200/200 | 3.96E+11 | 2.89E+12 | 1.02E+12 | d | d/c | 4.2 | 1.1 |

A cake deposit was formed at various flux rates (10, 20 and 200 L/m$^2$-h). The cake resistance at 20 L/m$^2$-h was ~28 %, and at 200 L/m$^2$-h 485 %, higher than at 10 L/m$^2$-h.

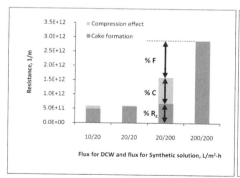 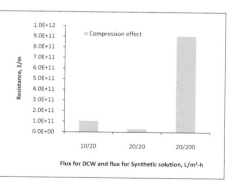

**Figure 6.18. Cake resistance and resistance due to flux effect- Canal water**

The resistance due to compression effect was significantly higher at 200 L/m$^2$-h than at 20 L/m$^2$-h; the increase was about 30 times.

By comparing the results for 20/200 and 200/200 it is possible to divide the effects that contribute to the total resistance in the cake formed at 200 L/m$^2$-

h: 24 % was due to "particles", 31 % was due to cake compression and 45 % was due to flux effect on arrangement of particles.

## 6.5.5   TIME EFFECT ON CAKE COMPRESSIBILITY

### 6.5.5.1   Seawater

In this case, the cake was initially formed at 20 L/m²-h and immediately after, the synthetic solution was filtered at 20, 80, 140 and 200 L/m²-h. The filtration of synthetic solution was allowed as long as possible to monitor a possible pressure increase on time.

The values of pressure and time during filtration of the synthetic solution are presented in Figure 6.19 for 20, 80, 140 and 200 L/m²-h. From 20 to 200 L/m²-h the pressure increased ~16 times to filter the same synthetic solution on the same built-up cake.

$$J = \frac{P}{\eta \cdot (R_m + R_c)}$$
                                                                        Eq.    6.25

As the synthetic solution has no fouling potential, $R_m + R_c$ correspond to the cake deposit initially formed at a low flux rate. An increase in flux directly produces an increase in pressure as presented in Eq. 6.10. This can be illustrated in Table 6.7.

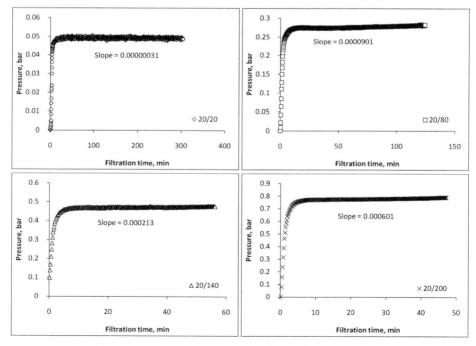

**Figure 6.19. Pressure development at various flux rates due to synthetic solution filtration on previously formed cake deposit at 20 L/m²-h**

**Table 6.7. Flux and pressure relation - Measured pressure values during filtration of synthetic solution**

| Flux | Ratio $J_i/J_{20}$ | $P_i/P_{20}$ | $P_{20}$, bar | $P_i$, bar | Measured $P_i/P_{20}$ | Difference |
|------|------|------|------|------|------|------|
| 20 | 1 | 1 | 0.051 | 0.052 | 1.03 | 3 % |
| 80 | 4 | 4 | 0.060 | 0.290 | 4.82 | 20 % |
| 140 | 7 | 7 | 0.053 | 0.507 | 9.41 | 34 % |
| 200 | 10 | 10 | 0.053 | 0.835 | 15.51 | 55 % |

The percentage difference in the last column can be attributed to the flux effect on the compression of the cake deposit producing a more compact cake (less porous) and in this way increasing the cake resistance and pressure.

At low flux (e.g., $< 20$ L/m²-h) a long filtration time is required for observing a time effect due to cake compression, especially to evaluate the influence in reverse osmosis systems.

### 6.5.5.2 Canal water

The cake formed at 20 L/m²-h was immediately placed under filtration with a synthetic solution at the same flux rate. The pressure values are plotted in Figure 6.20.

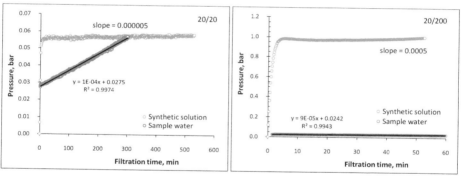

**Figure 6.20. Pressure development for filtration with synthetic solution at 20 and 200 L/m²-h**

At low flux filtration of synthetic solution in Figure 6.20: the pressure line for cake formation at 20 L/m²-h increased linearly over time, which means that the compression effect during cake formation can be ignored. The pressure line for filtration of synthetic solution at 20 L/m²-h appeared to increase very slightly over time (0.0005 %).

Over the period $50 - 100$ minutes, the slope of the pressure line increased very slightly and can be considered horizontal (0.0001%). The slope from $100 - 600$ minutes started to increase (0.0005%), and there was a time effect in the cake compression during the long filtration time even at very low flux.

At high flux (Figure 6.20, 20/200), the cake had already compressed since the beginning. The pressure values for cake formation at 20 L/m²-h increased linearly over time, which is similar to the previous test, and it can be assumed that the cake was not compressed during cake formation. The pressure increase for synthetic solution at 200 L/m²-h appeared to decrease very slightly and then gradually increased over time.

In the first period of filtration (up to 30 minutes) the slope was slightly negative (-0.0175 %). The slope started to increase after 40 minutes of filtration time and still continued afterwards (0.05 %).

**Table 6.8. Flux and pressure relation - Measured pressure values during filtration of synthetic solution**

| Flux | Ratio $J_i/J_{20}$ | Ratio $P_i/P_{20}$ | Pressure, bar | Ratio $P_i/P_{20}$ | Difference |
|------|--------------------|--------------------|---------------|--------------------|------------|
| 20   |                    |                    | 0.057         |                    |            |
| 200  | 10                 | 10                 | 0.975         | 17.1               | 71 %       |

The percentage difference in the last column (Table 6.8) - calculated from Eq. 6.10 - can be attributed to compression of the cake deposit producing a more compact cake (less porous) and in this way increasing the cake resistance and pressure. This difference is more significant with fresh water than with seawater.

Overall, the time effect was observed in the filtration of synthetic solution at low flux because the filtration can be performed for a longer time. On the other hand, with compression at high flux and short filtration time, the effect of compression was already apparent since the beginning, thus the time effect could not be obviously observed. If there was no limitation in the feed volume and cake compression at high flux can be performed over a longer period of time, then it is possible that the time effect can be observed also for high flux.

In practice, the consequences of time effect toward cake resistance might intensify the fouling occurring on the membrane elements. To measure the effect, firstly it is assumed that the RO cleaning frequency is 6 months (approximately equal to 259200 minutes). In addition, the applied flux is 20 L/m²-h and the fouling potential of the feed water at that flux can be ignored. Secondly, from the cake formation (at 20 L/m²-h) and cake compression (at 20 L/m²-h) experiments, it is known that the pressure increase due to time effect is averagely 0.0005 %/minute. Thus considering all the assumption above, if the time is taken into consideration in the fouling of RO membranes, the pressure increase within 6 months of operation will be about 1.3 %. However, it is expected that the pressure increase due to fouling will be higher as the feedwater still will have fouling potential to some extent.

## 6.5.6   ON-SITE MEASUREMENTS

In this section the results are presented of the on-site measurements with the MFI-UF constant flux set-up in three desalination pilot plants in Europe. The locations have been discussed previously.

## 6.5.6.1    Location A

The raw water sample was tested with 100 and 10 kDa membranes at various fluxes for the purpose of measuring the relation flux - MFI value. The results are presented in Table 6.9 and plotted in Figure 6.21.

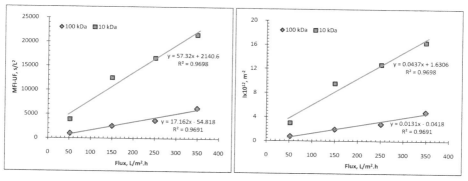

**Figure 6.21. I and MFI-UF values for RSW-A at various fluxes**

For the raw sea water, with 100 and 10 kDa membranes, the MFI-UF values as a function of flux showed a linear trend for the range of fluxes tested (50 – 350 L/m²-h). The regression coefficients are $R^2 = 0.97$ in both cases.

Based on the linear relations, the MFI-UF values at 15 L/m²-h (similar to SWRO operation) were projected. The projected value with the 10 kDa membrane is 15 times higher than with the 100 kDa membrane as shown in Table 6.9.

**Table 6.9. Raw seawater "A" MFI-UF values at various fluxes**

| Flux, L/m².h | 100 kDa, s/L² | 10 kDa, s/L² |
|---|---|---|
| 349 | 6300 | 21500 |
| 251 | 3700 | 16650 |
| 150 | 2550 | 12500 |
| 52 | 1000 | 3900 |
| 15* | 203 | 3000 |

\* Projected value from the linear equations in Figure 6.21.

The MFI-UF values with a 10 kDa PES membrane were measured for DMF effluent and for UF permeate at different fluxes. The results are shown in Table 6.10 and plotted in Figure 6.22.

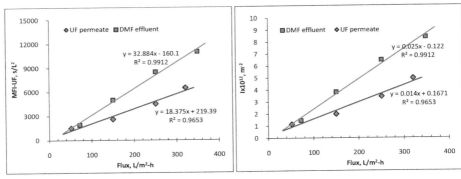

**Figure 6.22. I and MFI-UF values for Coag+DMF effluent and for UF permeate at various fluxes with 10 kDa membrane**

The MFI-UF values for DMF effluent are higher than the MFI-UF values for UF permeate. The obtained MFI-UF values fit a linear equation with good regression coefficients ($R^2$). In the case of UF permeate the $R^2$ was 0.97 and in the case of DMF effluent the $R^2$ was 0.99. The slopes of the equations are important to note, as the rate of MFI-UF change with flux is much higher with DMF effluent than with UF permeate (1.78 times).

With the obtained equations the MFI-UF value corresponding to a flux similar to SWRO systems operation (15 L/m²-h) was projected. The projections are presented in Table 6.10.

**Table 6.10. MFI-UF values for Coag+DMF effluent and for UF permeate with 10 kDa membrane**

| Flux, L/m²-h | Coag+DMF effl., s/L² | Flux, L/m²-h | UF permeate, s/L² |
|---|---|---|---|
| 349.3 | 11000 | 321 | 6500 |
| 251 | 8500 | 251 | 4500 |
| 150 | 5000 | 150 | 2600 |
| 72 | 1900 | 52 | 1500 |
| 15* | 333 | 15* | 495 |

\* Projected values from the linear equations in Figure 6.22.

Even though the measured MFI UF values for DMF effluent were higher than the UF permeate values, the projected MFI-UF values at 15 L/m²-h showed the opposite; DMF water's particulate fouling potential is lower than that of UF permeate. This may be attributed to the linear projection when obtaining the values at 15 L/m²-h as it is possible that at flux < 30 L/m²-h the fouling index may stabilize.

### 6.5.6.2    Location B

The raw water sample was tested with 30 and 10 kDa membranes at various fluxes for the purpose of measuring the relation flux - MFI value. The results are presented in Table 6.11 and plotted in Figure 6.23.

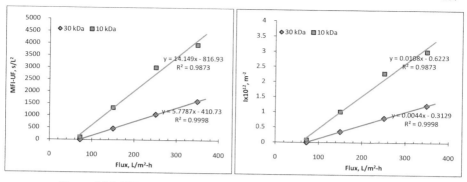

**Figure 6.23. MFI-UF and I values for Raw seawater "B" at various fluxes**

For the raw sea water, with 30 and 10 kDa membranes, the MFI-UF values as a function of flux showed a linear trend for the range of fluxes tested (72 – 350 L/m²-h). The regression coefficients ($R^2$) are above 0.98 in both cases.

**Table 6.11. Raw water MFI-UF values at various fluxes**

| Flux, L/m²-h | 30 kDa, s/L² | 10 kDa, s/L² |
|---|---|---|
| 349 | 1600 | 3950 |
| 251 | 1050 | 3000 |
| 150 | 460 | 1320 |
| 72.2 | 0 | 100 |
| 15* | 0 | 0 |

* Projected values from the linear equations in Figure 6.23.

Based on the linear relationships, the MFI-UF values at 15 L/m²-h (similar to SWRO operation) were projected. In this case, for both 30 and 10 kDa membranes the projection shows a zero MFI value, meaning the filtration resistance at this flux was minimal or below the detection limit of the test.

The MFI-UF values with a 10 kDa PES membrane were measured for UF-B permeate at different fluxes. The results are shown in Table 6.10 and plotted in Figure 6.22.

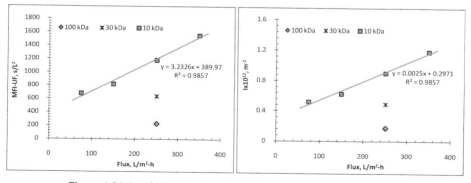

**Figure 6.24. I and MFI-UF values for UF-B permeate at various fluxes**

In addition to the 10 kDa results, the punctual values for 30 and 100 kDa (640 and 230 $s/L^2$ respectively) are included in Figure 6.24.

The obtained MFI-UF values for UF permeate with the 10 kDa membrane fit a linear equation with good regression coefficient ($R^2$). In the case of UF-B permeate the $R^2 = 0.98$. With the obtained regression equation, MFI $= 3.23 \times$Flux $+ 389$, the MFI value corresponding to a flux similar to SWRO systems operation (15 $L/m^2$-h) was projected. The projection is presented in Table 6.12.

Table 6.12. MFI-UF values for UF permeate at various fluxes with 10 kDa membrane

| Flux, $L/m^2$-h | UF permeate, $s/L^2$ |
|---|---|
| 349 | 1550 |
| 251 | 1180 |
| 150 | 820 |
| 75 | 680 |
| 15* | 438 |

* Projected value from the linear equation in Figure 6.22.

In comparison with the observed effect for the raw water in the previous section, for the UF-B permeate at 15 $L/m^2$-h the MFI-UF value is positive and around 440 $s/L^2$. This may be attributed to the narrower particle size distribution present in UF permeate that may create a less porous cake and therefore higher specific cake resistance.

### 6.5.6.3    Location C

The RO feed water sample was tested with 100, 50, 30 and 10 kDa membranes at various fluxes for the purpose of measuring the relation flux – I (MFI) value. The results are presented in Table 6.13 and plotted in Figure 6.25.

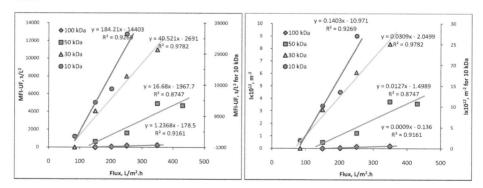

Figure 6.25. Flux effect measured on RO feed

For the RO feed water, $I$ and MFI-UF values as a function of flux showed a linear trend for the range of fluxes tested (72 – 400 $L/m^2$-h). The regression coefficients ($R^2$) are above 0.87 in all cases.

**Table 6.13. RO feed water MFI-UF values at various fluxes**

| Flux, L/m²-h | 100 kDa, s/L² | 50 kDa, s/L² | 30 kDa, s/L² | 10 kDa, s/L² |
|---|---|---|---|---|
| 350/430/350/250 | 235 | 4688 | 10959 | 35453 |
| 250/350/250/200 | 179 | 4906 | | 17686 |
| 200/250/150/150 | 50 | 1601 | 7981 | 13372 |
| 150/150/84/84 | 0 | 631 | 4058 | 2455 |
| 15* | 0 | 0 | 0 | 0 |

* Projected values from the linear equations in Figure 6.25.

Based on the linear relationships, the MFI-UF values at 15 L/m²-h (similar to SWRO operation) were projected. In all cases, for 100, 50, 30 and 10 kDa membranes the projections show a zero MFI value.

## 6.5.7   FOULING POTENTIAL

From the previous sections, the projected MFI-UF values at 15 L/m²-h can be summarized. Table 6.14 shows that in many cases for 30 and 100 kDa membranes it is not possible to measure a MFI-UF value (below detection limit or projection was a negative value).

**Table 6.14. Summary of MFI-UF values (s/L²) for various MWCO at 15 L/m²-h**

| Sample | 10 kDa | 30, kDa | 100, kDa |
|---|---|---|---|
| RSW-A | 3000 | - | 203 |
| CDMF-A | 333 | - | - |
| UF-A | 495 | - | - |
| RSW-B | bdl* | bdl | - |
| UF-B | 438 | - | bdl |
| RSW-C | - | - | |
| UF-C | bdl | bdl | bdl |

* bdl = below detection limit

Previous results indicate that the smaller the MWCO of the membrane used in the test, the larger the dependency of the fouling index on the flux rate ($m$ = slope).

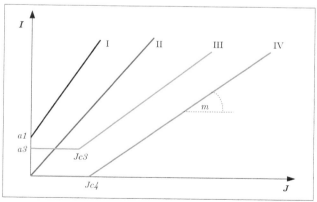

**Figure 6.26. Fouling index ($I$) versus Flux ($J$)**

As expressed in Eq. 6.14, the development of pressure $(\Delta P)$ in time depends on the flux $(J)$ value to the power two. From previous results we know that the fouling potential of water is also a function of flux as expressed in Table 6.15 and in reference to Figure 6.26. This suggests that the flux rate is even more significant than initially expected as fouling index is also a function of flux.

**Table 6.15. Fouling index ($I$) and pressure ($\Delta P$) as function of flux**

| Case | if: | $I =$ | $\Delta P =$ |
|------|-----|-------|--------------|
| I | | $a_1 + m \cdot J$ | $J \cdot \eta \cdot R_m + J^2 \cdot \eta \cdot (a_1 + m \cdot J) \cdot t$ |
| II | | $m \cdot J$ | $J \cdot \eta \cdot R_m + J^2 \cdot \eta \cdot (m \cdot J) \cdot t$ |
| III-A | $J < J_{c3}$ | $a_3$ | $J \cdot \eta \cdot R_m + J^2 \cdot \eta \cdot (a_3) \cdot t$ |
| III-B | $J > J_{c3}$ | $a_3 + m \cdot (J - J_{c3})$ | $J \cdot \eta \cdot R_m + J^2 \cdot \eta \cdot [a_3 + m \cdot (J - J_{c3})] \cdot t$ |
| IV-A | $J < J_{c4}$ | $0$ | $J \cdot \eta \cdot R_m$ |
| IV-B | $J > J_{c4}$ | $m \cdot (J - J_{c4})$ | $J \cdot \eta \cdot R_m + J^2 \cdot \eta \cdot [m \cdot (J - J_{c4})] \cdot t$ |

The coefficients "$a$" and "$m$" for real seawater samples are presented in Table 6.16 corresponding to the linear equations $(I = a + m \cdot J)$ observed in the previous sections for the sites A, B and C.

**Table 6.16. Summary of fouling index equations as a function of flux ($I = a + m \cdot J$)**

| | 10 kDa | | 30 kDa | | 100 kDa | |
|--------|--------|--------|--------|--------|---------|--------|
| Sample | a | m | a | m | a | m |
| RSW-A | 1.6306 | 0.0437 | - | - | -0.0418 | 0.0131 |
| CDMF-A | 0.122 | 0.025 | - | - | - | - |
| UF-A | 0.1671 | 0.014 | - | - | - | - |
| RSW-B | -0.6223 | 0.0108 | -0.3129 | 0.0044 | - | - |
| UF-B | 0.2971 | 0.0025 | - | - | - | - |
| RSW-C | | - | - | - | - | - |
| UF-C | -10.971 | 0.1403 | -2.0499 | 0.0309 | -0.136 | 0.0009 |

$I = 1/m^2$, $J = L/m^2\text{-h}$

The coefficients from the previous table were used to predict the pressure increase in time by using the equations in Table 6.15. In the projections were considered a similar flux to real RO operation ($\sim$15 L/m$^2$-h) and was assumed a particle deposition factor $\Omega = 1$. The projections for UF permeate in site B and C are presented in Figure 6.27..

According to Song *et al.* (2003) the membrane resistance $(R_m)$ in RO membranes varies from 1 to 1.5x10$^{11}$ Pa.s/m (1.5 being a conservative value) which is equivalent to a $R_m$ around 1 to 1.5x10$^{14}$ m$^{-1}$. The pressure increase described in Table 6.15 can be divided in two components: the first corresponding to the pressure required to overcome the resistance of the membrane $(J \cdot \eta \cdot R_m)$ which is assumed to be constant with the time; and the second component corresponding to the pressure increase due to the growth of the cake deposit on the surface of the membrane $(J^2 \cdot \eta \cdot (a + m \cdot J) \cdot t)$.

Two cases are presented in Figure 6.27. The one on the left shows a water from Site B where at low flux rates (10-30 L/m$^2$-h) the pressure increase due to cake resistance is high. The increase in pressure is illustrated in Figure 6.28 and it does not follow a power two relationship as suggested by Eq. 6.14 but a relationship as described by Table 6.15. The figure on the right shows the case of the site C, where at fluxes less than 75 L/m²-h no pressure increase is observed (fouling potential below detection limit). On Figure 6.27 should be included as well the pressure required to overcome the osmotic pressure (~31 bar for TDS = 35 g/L).

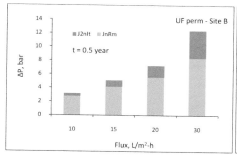 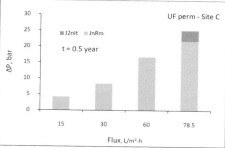

**Figure 6.27. Cake filtration pressure increase (ΔP) for RO feed water after 0.5 years as a function of flux. Calculated based on a 10 kDa membrane.**

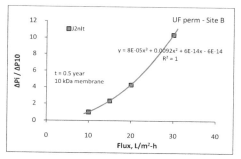

**Figure 6.28. Normalized pressure increase due to cake resistance at different fluxes for UF permeate - Site B**

Figure 6.28 shows the rate of pressure increase as a function of filtration flux after 6 months of operation with respect to the value at 10 L/m²-h. At 20 L/m²-h the pressure increase is 4.3 times that at 10 L/m²-h and at 30 L/m²-h the difference is 10.4 times, with respect to the reference value.

### 6.5.7.1   Particulate fouling potential at low flux rate

Typically a seawater reverse osmosis process system working at constant flux operates at around 15 L/m²-h. This is an average value per pressure vessel and not per element. Most of the RO pressure vessels contain 6 elements, where the NDP, flux and feed/concentrate/permeate concentrations are different for each element. The elements in the front produce the better permeate and higher flux in comparison to the elements at the rear. This is illustrated in the annex 6.8.3 where projections for a SWRO system are presented. The first element may produce ~20 L/m²-h but the last one will produce ~7.5 L/m²-h.

To further study the correlation of filtration flux and the potential of fouling of Delft canal water, several tests were performed at low flux rates i.e., 10, 20, 30, 40, and 50 L/m²-h. The results are shown in the Figure 6.29.

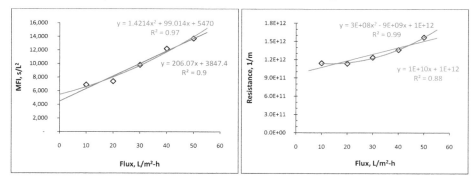

**Figure 6.29. MFI values (left) and Cake resistance (right) as a function of flux for the DCW cake layer (RC 100 kDa)**

MFI or I is more relevant since cake resistance depends on total filtered volume, which is arbitrary. However, in these tests the same volume of water was filtered.

From Figure 6.29, it is suggested that at low filtration rates the fouling potential of water does not decrease completely to zero but tends to level after around 20 L/m²-h.

In Figure 6.30 is illustrated an example of development of pressure (in one run, 60 min) in a UF system at different fluxes using the MFI-UF values from Figure 6.29.

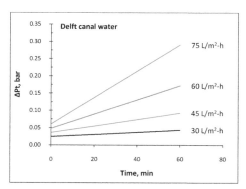

**Figure 6.30. Example of development of Pressure (in one run) in a UF system operating at different fluxes**

It can be observed that the development of pressure if highly dependent on the flux at which the UF system operates. After one hour operation, the pressure at 75, 60 and 45 L/m²-h is 6.7, ~4 and 2.14 times higher than the pressure at 30 L/m²-h.

## 6.6   Conclusions

A significant effect on fouling potential of the filtration flux was found.

Consequences of this effect are the following:

- In reverse osmosis systems, the fouling potential at low flux drops dramatically.
- In ultrafiltration systems, the rate of fouling increases at high fluxes in particular when flux > 60 L/m²-h.

This effect was observed due to:

- The effect of compression in the cake layer occurring even at low flux rates (e.g., 20 L/m²-h).
- The effect of flux on rearrangement of particles during cake formation occurring above a certain value. In case of the tested seawater, this value was around 60 L/m²-h.
- At low flux rates, the effect of flux is not clear.

The observed effect of flux in fouling potential has significant implications in fouling potential measurements like SDI, $MFI_{0.45}$ and MFI-UF constant flux. SDI and $MFI_{0.45}$ operate at constant pressure (2 bar) which yields high initial flux rates (> 1500 L/m²-h). As a consequence over estimation of fouling potential may occur.

MFI-UF constant flux can operate at any flux rate (10-350 L/m²-h). This is in advantage in considering the flux effect on fouling potential. To measure realistically the particulate fouling potential, test should be performed at same flux as RO systems (~20 L/m²-h) and MF/UF systems (60-80 L/m²-h).

## 6.7   List of abbreviations and symbols

### 6.7.1   ABBREVIATIONS

| | |
|---|---|
| DMF | Dual media filtration |
| kDa | Kilo Dalton |
| MFI-UF | Modified fouling index – ultra filtration |
| MWCO | Molecular weight cut off |
| PES | Polyethersulfone |
| RC | Regenerated cellulose |
| RO | Reverse osmosis |
| SDI | Silt density index |
| SWRO | Seawater reverse osmosis |
| UF | Ultra filtration |

## 6.7.2   SYMBOLS

| | |
|---|---|
| $A$ | Effective membrane surface area (m²) |
| $A_o$ | Standard reference area of the MFI 0.45 μm membrane (13.8×10⁻⁴) (m²) |
| $C_b$ | Concentration of particles in a feed water (kg/m³) |
| $d_p$ | Diameter of particles forming the cake (m) |
| $I$ | Fouling index of particles in water to form a layer with hydraulic resis. (m⁻²) |
| $J$ | Permeate water flux (m³/m²·s) or (L/m²·h) |
| $K_w$ | Permeability constant for water (m³/m²·s·bar) |
| $R_c$ | Cake formation resistance (m⁻¹) |
| $R_m$ | Membrane resistance (m⁻¹) |
| $r_p$ | Pore radius (m) |
| $R_t$ | Total resistance (m⁻¹) |
| $T$ | Filtration time (second) |
| $T$ | Temperature of feed water (°C) |
| $V$ | Filtrate volume (m³) |
| $P$ | Applied trans-membrane pressure (bar or N/m²) |
| $P_o$ | Standard reference applied trans-membrane pressure (bar or N/m²) |
| $P_c$ | Pressure drop over the cake (bar or N/m²) |
| $x$ | Membrane thickness (m) |
| $\alpha$ | (Average) specific cake resistance (m/kg) |
| $\alpha_o$ | Initial specific cake resistance (m/kg) |
| $\Omega$ | Deposition factor (-) |
| $\varepsilon$ | Cake / membrane surface porosity (-) |
| $\eta_{20\,C}$ | Water viscosity at 20 °C (N·s/m²) |
| $\eta_T$ | Water viscosity at temperature T (N·s/m²) |
| $\rho_p$ | Density of particles forming the cake (kg/m³) |
| $\tau$ | Tortuosity of membrane pores |
| $\omega$ | Compressibility coefficient (-) |
| $\psi$ | Cake ratio (-) |
| $\Delta P_T$ | Total applied pressure drop (N/m²) |
| $\Delta P_m$ | Pressure drop across membrane (N/m²) |
| $\Delta P_C$ | Pressure drop across the cake layer (N/m²) |

# 6.8   Annex

## 6.8.1   COMPRESSION WITH ULTRA PURE WATER

**Table 6.17. Filtration resistances for clean membrane, cake deposit and compressed cake deposit - Seawater**

| MWCO, kDa | Formation / Compression flux rates | Rm, 1/m | Rcf, 1/m | Rcc, 1/m | Ratios id | Ratio | Cake formation | Cake compression |
|---|---|---|---|---|---|---|---|---|
| 50 | 20/20 | 6.52E+11 | 4.06E+11 | -1.51E+10 | a | | | |
| | 20/200 | 6.49E+11 | 3.69E+11 | 1.37E+11 | b | b/a | 0.9 | - |
| | 200/200 | 6.42E+11 | 1.20E+12 | -3.22E+11 | c | c/b | 3.2 | - |

When using ultra pure water (UPW), instead of increasing the cake resistance the opposite effect was observed; this might be related to the big difference in ionic strength between the formed cake with seawater and the solution used for filtration.

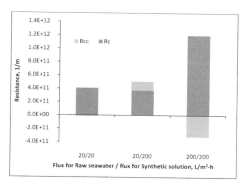

**Figure 6.31. Cake formation resistance and cake compression resistance – Seawater and 50 kDa**

Rc means resistance during cake formation and Rcc means resistance after filtering UPW.

## 6.8.2    SYNTHETIC SOLUTION

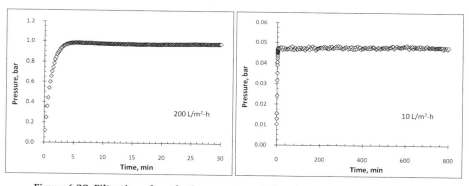

**Figure 6.32. Filtration of synthetic seawater solution through a 10 kDa membrane**

## 6.8.3    SWRO DESIGN PROJECTIONS

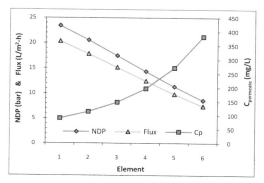

**Figure 6.33. NDP, flux and permeate concentration projections for a 15 m³/h SWRO system using SWC6 elements**

# 6.9    References

ALMY, C. & LEWIS, W. K. 1912. Factors Determining the Capacity of a Filter Press. *Journal of Industrial & Engineering Chemistry*, 4, 528-532.

ALTHULUTH, M. 2009. *Further Development of the Modified Fouling Index (MFI-UF) at Constant Flux for SWRO Applications*. Master, UNESCO-IHE.

BOERLAGE, S. F. E. 2001. *Scaling and Particulate Fouling in Membrane Filtration Systems*, Lisse, Swets&Zeitlinger Publishers.

BOERLAGE, S. F. E., KENNEDY, M., TARAWNEH, Z., FABER, R. D. & SCHIPPERS, J. C. 2004. Development of the MFI-UF in constant flux filtration. *Desalination*, 161, 103-113.

BOERLAGE, S. F. E., KENNEDY, M. D., ANIYE, M. P., ABOGREAN, E. M., GALJAARD, G. & SCHIPPERS, J. C. 1998. Monitoring particulate fouling in membrane systems. *Desalination*, 118, 131-142.

BOERLAGE, S. F. E., KENNEDY, M. D., DICKSON, M. R., EL-HODALI, D. E. Y. & SCHIPPERS, J. C. 2002. The modified fouling index using ultrafiltration membranes (MFI-UF): characterisation, filtration mechanisms and proposed reference membrane. *Journal of Membrane Science*, 197, 1-21.

CARMAN, P. C. 1938. Fundamental principles of industrial filtration (A critical review of present knowledge). *Trans. Instn Chem. Engrs*, 16, 168-188.

CHERYAN, M. 1998. *Ultrafiltration and Microfiltration Handbook*, Lancaster, USA.

CHOI, S. W., YOON, J. Y., HAAM, S., JUNG, J. K., KIM, J. H. & KIM, W. S. 2000. Modeling of the permeate flux during microfiltration of BSA adsorbed microspheres in a stirred cell. *Journal Colloid Interface Science*, 228, 270-278.

COULSON, J. M. & RICHARDSON, J. F. 1990. *Chemical Engineering Volume Two: Particle Technology & Separation Processes*, Butterworth-Heinemann.

GUIGUI, C., ROUCH, J. C., DURAND-BOURLIER, L., BONNELYE, V. & APTEL, P. 2002. Impact of coagulation conditions on the in-line coagulation/UF process for drinking water production. *Desalination*, 147, 95-100.

HERMIA, J. 1982. Constant pressure blocking filtration laws - Application to power-law non-newtonian fluids. *Trans IChemE*, 60, 183-187.

KOVALSKY, P., WANG, X., BUSHELL, G. & WAITE, T. D. 2008. Application of local material properties to prediction of constant flux filtration behaviour of compressible matter. *Journal of Membrane Science*, 318, 191-200.

RIETEMA, K. 1953. Stabilizing effects in compressible filter cakes. *Chemical Engineering Science*, 2, 88-94.

RUTH, B. F. 1935. Studies in Filtration III. Derivation of General Filtration Equations. *Industrial & Engineering Chemistry*, 27, 708-723.

SANTIWONG, S. R. 2008. *Analysis of compressible cake behaviour in submerged membrane filtration for water treatment.* PhD Dissertation, UNSW.

SCHIPPERS, J. C. 1989. *Vervuiling van hyperfiltratiemembranen en verstopping van infiltratieputten,* Rijswijk, Keuringinstituut voor waterleidingartikelen KIWA N.V.

SONG, L., HU, J. Y., ONG, S. L., NG, W. J., ELIMELECH, M. & WILF, M. 2003. Performance limitation of the full-scale reverse osmosis process. *Journal of Membrane Science,* 214, 239-244.

TILLER, F. M. & COOPER, H. R. 1960. The Role of Porosity in Filtration: IV. Constant Pressure Filtration. *A.I.Ch.E. Journal,* 6, 595-601.

# Chapter 7

# 7 Particle deposition in SWRO systems

Chapter 7 is based on:

SALINAS RODRÍGUEZ, S. G., KENNEDY, M. D., AMY, G. & SCHIPPERS, J. C. (2011). Particle deposition in SWRO systems. *Water Research*, submitted.

SALINAS RODRÍGUEZ, S. G., KENNEDY, M. D., AMY, G. & SCHIPPERS, J. C. (2010). Flux dependency of particulate fouling by MFI-UF measurements in seawater reverse osmosis systems. In: EDS (ed.) *EuroMed 2010: Desalination for Clean Water and Energy*. Tel-Aviv, Israel: EDS.

# 7.1   Introduction

Particles are present in all waters. In membrane filtration, when particles are too large to enter the membrane pores, a sieving process occurs. The retained particles accumulate on the membrane surface in a growing cake layer. In a cross flow filtration mode, the fluid motion tangential to the membrane surface may arrest the cake growth so that extended operation is possible. Unlike dead-end filtration, the cake layer does not grow indefinitely (until backwash or cleaning occurs); instead, the high shear exerted by the water flowing tangentially to the membrane surface sweeps the particles toward the end of the membrane element so that the cake layer remains relatively thin (Belfort et al., 1994). Reverse osmosis sytems operate in a cross flow fashion.

The amount of particles accumulating on the membrane surface influence the operation of the system by increasing the feed pressure to maintain constant productivity, or by decreasing the water production at constant pressure operation. Also, to predict the rate of particulate fouling in RO systems, it is necessary to consider the rate of particles depositing on the membranes. The rate of deposition can be measured by the deposition factor.

In this chapter the amount of particles present in the water is measured with the modified fouling index (MFI) at constant flux. Measurments are performed in the feed and concentrate waters of two seawater reverse osmosis (SWRO) systems. Different membrane molecular weight cut-offs (MWCO) are used in parallel and in serial filtration.

As RO membranes also reject ions besides particles, the salinity of the feed water increases inside the pressure vessel. The extent of increase in salinity is governed by the recovery of the system. For this reason, the effect of increasing salinity from the RO feed to RO concentrate was studied.

The deposition factor measurements will be applied in the next chapter to predict the rate of particulate fouling in several locations.

# 7.2   Background

## 7.2.1   PARTICLES

There are two general types of particles in natural waters, hydrophobic (water repelling) and hydrophilic (water attracting). Hydrophobic particulates have a well-defined interface between the water and solid phases and have a low affinity for water molecules. In addition, hydrophobic particles are thermodynamically un-stable and will aggregate irreversibly over time (Crittenden et al., 2005).

Hydrophilic particulates such as clays, metal hydroxides, proteins, or humic acids have polar or ionized surface functional groups. Many inorganic particles in natural waters, including hydrated metal oxides (iron or aluminium oxides), silica ($SiO_2$), and asbestos fibres, are hydrophilic because water molecules will bind to the polar or ionized surface functional groups (Stumm and Morgan, 1996). Many organic particulates are also hydrophilic and include a wide diversity of bio-colloids (humic acids, viruses) and suspended living or dead microorganisms (bacteria, protozoa, algae). Because bio-colloids can adsorb on the surfaces of inorganic particulates, the particles in natural water often exhibit heterogeneous surface properties. Some particulate suspensions such as humic or fulvic acids may be reconstituted after aggregation and are reversible because they are bonded together by hydrogen bonding.

In nature, most colloids and particles are negatively charged. This knowledge has been used by membrane manufacturers to influence the surface charge of the membranes so as to repel suspended particles (Belfort et al., 1994).

Figure 7.1 shows the division between dissolved and particulate organic carbon, based on filtration through a 0.45 µm filter. Nevertheless, overlapping the dissolved and particulate fractions is the colloidal fraction. According to IUPAC (1971), the term colloidal refers to a state of subdivision, implying that the molecules or poly-molecular particles dispersed in a medium have at least in one direction a dimension of roughly between 0.001 µm and 1 µm, or that in system discontinuities are found at distances of that order. Therefore, colloids have a size between 0.001 µm and 1 µm. Figure 7.1 indicates an equivalence of 10 kDa for 0.001 µm.

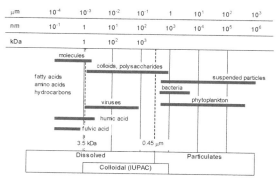

**Figure 7.1. Continuum of particles, colloids and dissolved organic carbon in natural waters (Aiken and Leenheer, 1993)**

A system containing colloidal particles is said to be stable if during the period of observation, it is slow in changing its state of dispersion (Crittenden et al., 2005).

## 7.2.2   DEPOSITION FACTOR

Only a fraction of the RO feed water is forced to pass through the membranes. This fraction of water depends on the recovery ($R$) at which the

RO unit operates. In dead-end filtration all the particles bigger than the membrane's pores will be retained while in the case of cross-flow, only the fraction of water passing through the membranes is affected and the associated fraction of particles may or may not be accumulated on the membrane surface.

The deposition factor was first proposed by Schippers et al. (1980, 1981) in a model to predict flux decline in reverse osmosis systems. It was defined as the fraction of particles deposited, which are present in the water passing the reverse osmosis membrane.

A few years later, Schippers (1989) presented the results obtained in a pilot plant working with water from the IJsselmeer lake located in the north of The Netherlands. The total recovery of the installation was 90 % in four stages. The deposition factor was obtained by measuring at constant pressure the MFI (0.05 μm) values of the RO feed and RO concentrate waters. The majority of measured values in the four stages were less than one, meaning that only a fraction of the fed particles attached to the membranes. Some values were negative meaning that some particles were separated from the membrane.

Boerlage et al. (2001, 2003) presented the results of measurements with MFI-UF constant pressure in two locations working with fresh water from the river Rhine and from the IJssel lake. The deposition factor values for the IJssel lake plant and for the river Rhine plant were all negative. The results were attributed to changes in the composition of the cake formed on the RO membranes over time due to the forces acting on the particle in tangential flow.

Recently, Sioutopoulus et al. (2010) worked with colloidal organic and inorganic species to link fouling potential between UF and RO. The experimental set-up was a bench scale RO unit (SEPA cell type). The salinity levels in the water were 500, 2000, 5000 and 10000 mg/L as TDS. The range of fluxes tested was 25-40 $L/m^2$-h with a water recovery of 1-2 %. In this study, the deposition factor was obtained by measuring the ratio of actual fouling species deposited on the membrane over the theoretical one. The author mentioned that the theoretical mass deposition values were calculated based on the total permeate volume of each RO test. Thus, the mean deposition factor values were estimated to be 0.6, 0.9 and 1.0 for humic acids, sodium alginate and ferric oxide, respectively.

Many studies have been conducted to understand factors affecting fouling of membranes. Results of membrane autopsies illustrate that biofouling and organic fouling may occur preferably in the first element while precipitation of salts (scaling) is expected to occur in the last elements. Furthermore, the fouling layer distribution may not be homogenous over the entire membrane surface.

In a RO pressure vessel, the flux distribution along the vessel is not uniform; the front elements have a higher production rate in comparison with the rear elements (illustrated in annex 7.9.1) that have a lower production rate. Furthermore, the cross flow velocity in the front and rear elements are not uniform.

In some cases, inside the pressure vessel, the placed elements are not identical. It is possible to observe that in the front of the pressure vessel are placed high production membranes while at the end high rejection membranes. This results in a situation where the water towards the membrane is non equally distributed.

Many studies have focused on the effect of the channel geometry, and shear rate on colloidal fouling in cross flow (Hoek et al., 2002).

All of these factors (non uniform flux rate, cross flow velocities, geometry of spacer) make it difficult to study the deposition of particles in RO units. In case of any measurements, they have to be performed on site and considering retention times to test the same water if possible and the measured values are an average of the RO pressure vessel.

Furthermore, it is possible that a preferential deposition of particles may occur and influence the measurements of particulate deposition through MFI. In this case the size distribution of particles in the feed water may differ from the particle size distribution in the concentrate water. Current methods to measure particles size (e.g., laser diffraction, microscopy) are limited in working at levels less than 0.05 µm.

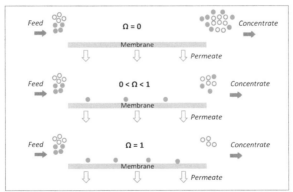

**Figure 7.2. Particle deposition in cross flow filtration on permeable surfaces**

Figure 7.2 shows schematically the particle deposition on a membrane surface considering 50 % recovery. Empty circles are the fraction of particles that are not accumulated on the membrane; and full circles are the fraction of particles that might be accumulated on the membrane.

Particle's size plays a role in particle deposition on permeable surfaces; the deposition rate has been assumed to be lower for larger particles compared to smaller particles. This is due to the fact that the back transport by inertial lift is significant for larger particles (Song and Elimelech, 1995). Chellam and Wiesner (1998) reported that the cake formed in cross flow mode had a higher percentage of fine particles resulting in a higher specific cake resistance compared to the feed suspension.

In this sense, it is important to accurately measure the amount of particles that are entering the RO unit, as well as how much particles are leaving the plant in the permeate water and in the concentrate water. This can be performed by doing a "mass" balance.

## 7.2.3   MASS BALANCE EQUATIONS

A schematic of a RO unit is presented in Figure 7.3. From this, a flow balance and mass balance can be performed as in Eq. 7.1 and Eq. 7.2, respectively.

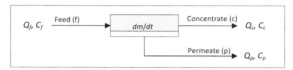

**Figure 7.3. RO membrane schematic**

$$Q_f = Q_p + Q_c \qquad\qquad\qquad\qquad\qquad\qquad \text{Eq.} \quad 7.1$$

$$Q_f \cdot C_f = Q_p \cdot C_p + Q_c \cdot C_c \qquad\qquad\qquad\qquad\qquad \text{Eq.} \quad 7.2$$

In these equations, it is important to notice that only the permeate water $(Q_p)$ has passed through the membrane and therefore it is only from this volume of water that the membrane is rejecting ions, organic matter and particles. The rest of the water (concentrate) passes tangentially to the membrane without any change.

To consider the particles being accumulated on the membrane, the term $dm/dt$ is introduced in the mass balance. Then we have:

$$Q_f \cdot C_f = Q_p \cdot C_p + Q_c \cdot C_c + \frac{dm}{dt} \qquad\qquad\qquad \text{Eq.} \quad 7.3$$

In RO systems, only a part of the feed water passes through the membranes $(Q_p)$. The extent of water passing through the membrane elements depends on the recovery of the system $(R)$. From the part of water that passes through the membranes and where all the particles are rejected, only a fraction will be accumulated $(\Omega \cdot C_f)$ on the surface and the other part will appear in the concentrate. So the part of particles that accumulate on the surface of the membrane can be expressed as $\Omega \cdot C_f \cdot Q_p$. Then,

$$Q_f \cdot C_f = Q_p \cdot C_p + Q_c \cdot C_c + \Omega \cdot Q_p \cdot C_f \qquad\qquad \text{Eq.} \quad 7.4$$

Assuming the particle concentration in the permeate is zero (100 % rejection), therefore $C_p = 0$. Consequently,

$$Q_f \cdot C_f = Q_c \cdot C_c + \Omega \cdot Q_p \cdot C_f \qquad \text{Eq.} \quad 7.5$$

Rearranging the previous equation, we have:

$$\Omega \cdot Q_p \cdot C_f = Q_f \cdot C_f - Q_c \cdot C_c \qquad \text{Eq.} \quad 7.6$$

Then,

$$\Omega = \frac{Q_f \cdot C_f - Q_c \cdot C_c}{Q_p \cdot C_f} \qquad \text{Eq.} \quad 7.7$$

Rearranging Eq. 7.7,

$$\Omega = \frac{Q_f \cdot C_f}{Q_p \cdot C_f} - \frac{Q_c \cdot C_c}{Q_p \cdot C_f} \qquad \text{Eq.} \quad 7.8$$

$$\Omega = \frac{Q_f}{Q_p} - \frac{Q_c}{Q_p} \cdot \frac{C_c}{C_f} \qquad \text{Eq.} \quad 7.9$$

On the other hand, the system recovery (R) is defined as:

$$R = \frac{Q_p}{Q_f} \cdot 100 \qquad \text{Eq.} \quad 7.10$$

Rearranging the previous equation, we have:

$$\frac{Q_f}{Q_p} = \frac{1}{R} \qquad \text{Eq.} \quad 7.11$$

From Eq. 7.1, the concentrate flow is:

$$Q_c = Q_f - Q_p \qquad \text{Eq.} \quad 7.12$$

Then,

$$\frac{Q_c}{Q_p} = \frac{(Q_f - Q_p)}{Q_p} = \frac{Q_f}{Q_p} - 1 \qquad \text{Eq.} \quad 7.13$$

Replacing Eq. 7.11 and Eq. 7.13 in Eq. 7.9,

$$\Omega = \frac{1}{R} - \left(\frac{Q_f}{Q_p} - 1\right) \cdot \left(\frac{C_c}{C_f}\right) \qquad \text{Eq.} \quad 7.14$$

Replacing Eq. 7.11 in Eq. 7.14,

$$\Omega = \frac{1}{R} - \left(\frac{1}{R} - 1\right) \cdot \left(\frac{C_c}{C_f}\right) \qquad \text{Eq.} \quad 7.15$$

and rearranging Eq. 7.15 we have,

$$\Omega = \frac{1}{R} - \left(1 - \frac{1}{R}\right) \cdot \left(\frac{C_c}{C_f}\right)$$                                      Eq.   7.16

Then, we can obtain the deposition factor equation as function of recovery,

$$\Omega = \frac{1}{R} + \frac{C_c}{C_f} \cdot \left(1 - \frac{1}{R}\right)$$                                      Eq.   7.17

Or as function of concentration factor $(CF)$,

$$\Omega = \frac{1}{(CF - 1)} \cdot \left(CF - \frac{C_c}{C_f}\right)$$                                      Eq.   7.18

Where the concentration factor is:

$$CF = \frac{1}{1 - R}$$                                      Eq.   7.19

The formula above assumes that the particle rejection is 100 %.

In this study, $C_f$ and $C_c$ correspond to $MFI_{feed}$ and $MFI_{concentrate}$. Equations 7.17 and 7.18 are illustrated in Figure 7.4.

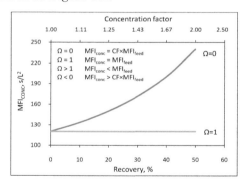

**Figure 7.4. Deposition factor as function of RO recovery and of conversion factor**

A *positive* deposition factor indicates particles are being accumulated on the membrane surface as they pass through the system while a *negative* factor indicates the number of particles in the concentrate exceeds the incoming flux (taking into account the concentration factor) (Boerlage, 2001, Schippers, 1989).

There are possible scenarios from previous equations:

- $\Omega = 0$ means $C_c = C_f \times CF$     No particles deposit
- $\Omega = 1$ means $C_c = C_f$     All particles deposit
- $\Omega > 1$ means $C_c < C_f$     All particles deposited
         + retention inside pressure vessel (e.g., spacer)

- $\Omega < 0$ means $C_c > C_f \times CF$          Particles might be removed inside pressure vessel; earlier deposited particles released; particles formed by bacteria; particle size distribution influence results.

## 7.2.4   PARTICLE DEPOSITION MECHANISMS

When particles enter the feed channel in the membrane element and get close to the membrane surface, two forces are imposed on particles namely: *i)* convective force towards the membrane surface (due to the drag force of permeation flow) and *ii)* the shear force (due to crossflow velocity).

The particle backtransport mechanisms include concentration polarisation (brownian diffusion, influencing small colloids), shear induced diffusion and inertial lift (influencing big particles) (Belfort et al., 1994). In recent studies, it was reported that particle-particle and particle-membrane interactions (including entropy, van der Waals interactions and electrostatic interactions) may also play important roles in particle transport to and/or from the membrane surface, especially in concentrated solutions of colloidal particles (Davis, 1992, Jiang, 2007).

The random movement resulting from the bombardment of particles by water molecules is defined as brownian diffusion. Shear induced diffusion occurs when individual particles undergo random displacements from the stream lines in a shear flow as they interact with and tumble over other particles (Davis and Sherwood, 1990). Belfort et al. (1994) mentioned that the back-diffusion of particles away from the membrane is supplemented by a lateral migration of particles due to inertial lift (also known as tubular-pinch effect).

The three backtransport mechanisms work simultaneously, and the total backtransport velocity is assumed to be the sum of them (Jiang, 2007). The contribution of the individual mechanisms depends on the particle size and crossflow velocity.

## 7.3   Objectives

The objectives of this chapter are the following:

- Study the salinity effect when measuring the MFI-UF value for RO feed and for RO concentrate.
- Measure the particle deposition factor in SWRO systems.
- Measure the effect of membrane MWCO on the deposition factor.
- Measure particle size distribution by fractionating the RO feed and RO concentrate water. Fractionation in parallel and fractionation in series are performed.

## 7.4   Material and methods

### 7.4.1   MEMBRANES

Various MWCO membranes; 100, 50, 30 and 10 kDa polyethersulfone (PES), 25 mm diameter membrane filters; were used in the tests. In advance of the on-site testing, all of the membranes were first cleaned and the membrane resistance ($R_m$) was measured with ultra pure water (UPW).

The average $R_m$ values per batch of membranes and standard deviation are presented on Table 7.1. The manufacturer (Millipore) provides packages of 10 membranes. The outlier membranes with $R_m$ values higher or lower than standard deviation value were not used for the testing.

**Table 7.1. Average R$_m$ for the PES membranes used in the testing**

|  | Complete set(s) | | # total/ | Partial set(s) | |
|---|---|---|---|---|---|
| Membranes | $R_m$, m$^{-1}$ | Std Dev, % | #excluded* | $R_m$, m$^{-1}$ | Std Dev, % |
| 100 kDa PES | $3.21 \times 10^{11}$ | 9 % | 10/2 | $3.10 \times 10^{11}$ | 5 % |
| 50 kDa PES | $6.59 \times 10^{11}$ | 5.4 % | 8/2 | $6.58 \times 10^{11}$ | 3.3 % |
| 30 kDa PES | $7.57 \times 10^{11}$ | 5.9 % | 17/3 | $7.48 \times 10^{11}$ | 3.4 % |
| 10 kDa PES | $1.03 \times 10^{12}$ | 8 % | 20/5 | $9.97 \times 10^{11}$ | 5 % |

*Outliers were excluded

The membrane filters were each transported to the testing location soaked in 50 ml UPW.

### 7.4.2   CONSTANT FLUX FILTRATION SET-UP

The filtration set-up has been described in detail in chapter 5. A schematic of it is presented in Figure 7.5.

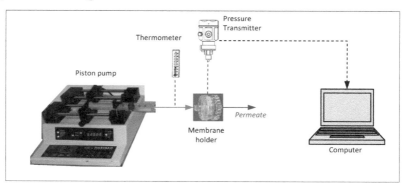

**Figure 7.5. Constant flux filtration set-up**

The set-up was used to measure the Modified Fouling Index - Ultrafiltration (MFI-UF) of the RO feed and of the RO concentrate at constant flux (250 L/m$^2$-h) with various membrane MWCOs (e.g., 100, 50, 30, and 10 kDa).

The set-up is portable enough to be transported for on-site testing in the chosen facilities.

## 7.4.3 ON SITE TESTING LOCATIONS

Two locations were selected for the tests. One is located in the North Sea and the other one is located on the Mediterranean Sea. The locations and pre-treatments are briefly described on Table 7.2.

**Table 7.2. Description of the tested locations**

| Location | Intake | Pre-treatment | RO unit |
|---|---|---|---|
| A (North-Western Mediterranean water) | Submerged pipe (L = 2.5 km) | UF (0.02 µm) | R = 45 %. J = 15 L/m²-h. 7 elements/vessel. 8" modules. |
| B (North Sea water) | Submerged pipe (L = 100 m) | Strainer – UF (~300 kDa) | R = 40 %. J = 15 L/m²-h. 6 elements/vessel. 8" modules. |

The water samples (RO feed water and RO concentrate water) were taken time-lagged, considering the hydraulic residence time in the RO pressure vessel. A summary of the water qualities is presented on Table 7.3. Location B has almost twice DOC concentration in comparison with location A.

**Table 7.3. Summary of water characteristics**

| Sample | Location | DOC, mg/L | pH | T, °C | SUVA (L/mg·m) | EC, mS/cm | NTU |
|---|---|---|---|---|---|---|---|
| Raw water | A | 0.75 | 8.1 | 16.4 | 0.55 | 57-58 | 0.5-1.5 |
| UF perm / RO feed | A | 0.72 | 8.0 | 16.4 | 0.45 | 57-58 | |
| RO conc | A | 1.5 | | | 0.5 | | |
| Raw water | B | 1.45 | 8.1 | 13.5 | 2.3 | 48.5 | 8-12 |
| UF perm / RO feed | B | 1.3 | 6.5 | 13.5 | 2.15 | 48.5 | |
| RO conc | B | 2.2 | | | 2.0 | 76.5 | |

To reduce the impact of variations in water quality, as many tests as possible were performed during the same day.

## 7.4.4 PARALLEL AND IN-SERIES MEASUREMENTS

The MFI-UF values for RO feed and for RO concentrate were measured at various MWCOs. The deposition factor was calculated according to Eq. 7.17. Figure 7.6 shows the schematic of the performed tests in parallel.

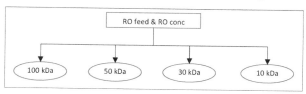

**Figure 7.6. Schematic of the "in parallel" filtration tests**

For the fractionation in series, the schematic in Figure 7.7 illustrates the followed procedure. In this case the particles are compared by size range in the feed and concentrate of the RO.

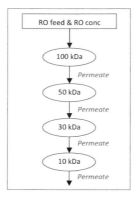

**Figure 7.7. Schematic of the "in series" filtration tests**

Enough volume of water had to be produced by the 100 kDa membrane to ultimately reach the 10 kDa filter.

## 7.5   Results and discussion

Duplicates and in some cases triplicates of results have been performed in all tests and the average value is presented in the following sections.

### 7.5.1   SALINITY EFFECT

In reverse osmosis as water passes through the membranes, the salinity and the particle concentration in the water increase. This increase is a function of the recovery or conversion of the RO. For instance, a RO system working at 50 % recovery will have the concentrate concentration twice the feed concentration. This salinity increase is illustrated in the annex 7.9.2 where projections for a 40 % recovery SWRO system are presented.

#### 7.5.1.1   Canal water + NaCl solution

The increase in salinity in the RO concentrate due to recovery (concentration factor) may influence the MFI-UF values. For this reason, the effect of salinity on various solutions was studied and measured with the MFI-UF at constant flux. A solution was prepared by dissolving NaCl salt (99.9999 %) in ultra pure water (UPW) and stirring at high speed and heating the solution to 40° C to dissolve NaCl completely; then the solution was allowed to cool down to room temperature. Then, the solution was mixed with Delft canal water (DCW), which was previously 0.45 μm pre-filtered, to achieve a 50 % mixing. The tests were performed with two different DCW batches (March and April 2009).

The results shown in Figure 7.8 indicate that the MFI–UF values increased with increasing salinity of the solution. The same trend was observed for the two batches of DCW.

For the first batch of DCW, the increase in MFI value from 0 addition to 100,000 mg/L was 325 % compared with ~70 % for the second batch. Comparing the increase in the same range of salinity as seawater (35,000 mg/L and recovery 40 %), for the first batch the increase was about 60 % while for the second batch the increase in MFI value was about 10 %.

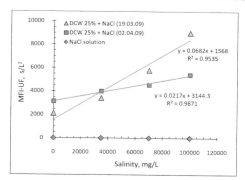

**Figure 7.8. MFI-UF of DCW 25 % as function of ionic strength of NaCl (100 kDa RC, 250 L/m²-h)**

Mendret et al. (2009) observed a decrease in permeate flux in cross flow filtration with increasing the addition of $KNO_3$ over the range of $1 \times 10^{-5}$ M to $1 \times 10^{-1}$ M (0.1 to 10,000 mg/L) due to a reduction in porosity of the cake formed. A different trend was observed by Boerlage (2001, 2003) where the MFI-UF values increased with increasing salinity up to a maximum value at 6,000 mg/L; after this point the MFI-UF decreased, however, it was only tested up to 10,000 mg/L.

The above trends can be explained by that fact the particles and surfaces in a polar medium such as water are often electrically charged and surrounded by a double layer (Song and Elimelech, 1995). An increase in ionic strength is expected to compress the double layer around the particles and reduce their inter-particles distance. Consequently, the cake layer may be more densely packed and the resistance to permeate flow is expected to increase. For the inter-particle distance, the Debye screening length, $K^{-1}$, is given by:

$$K^{-1} = \left( \frac{2 \times 10^3 \cdot e^2 \cdot N_A \cdot IS}{\varepsilon_r \cdot \varepsilon_0 \cdot k \cdot T} \right)^{-1/2}$$

Eq.    7.20

Where: $e$ is the elementary charge, $N_A$ is Avogadro's number, $IS$ is the ionic strength, $\varepsilon_r$ is the dielectric permittivity of water, $\varepsilon_0$ is the permittivity of free space, $k$ is the Boltzmann constant, and $T$ is the absolute temperature. Eq. 7.20 indicates that when the ionic strength increases, the range of the repulsive force between the colloidal particles decreases (Faibish et al., 1998). Similarly, the double layer interaction between particles and membrane surface is influenced by ionic strength, and deposition rate increases with increasing solution ionic strength. This is attributed to a decrease in the repulsive double layer forces between the particles and surfaces as the ionic strength increases (Song and Elimelech, 1995).

### 7.5.1.2   RO feed seawater + NaCl solution

A solution was prepared by dissolving NaCl salt (99.9999 %) in UPW, then the solution was stirred at high speed and heated to 40° C to dissolve NaCl completely, then cooled down to room temperature. After that, the solution was mixed with RO feed seawater at low intensity.

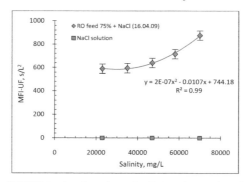

**Figure 7.9. MFI-UF of SWRO feed as function of salinity - Site B. (100 kDa RC, 250 L/m²-h)**

In Figure 7.9, the MFI–UF of RO feed (75 % SW mixed with 25 % UPW) without salt addition was 590 s/L². As the ionic strength increased from 23,000 to 47,000 mg/L, the increase in MFI-UF value was not significant (~8 %). When ionic strength increased from 47,000 to 70,000 mg/L the MFI-UF increased by ~35 % (from 640 to 875 s/L²).

From Figure 7.9 when ionic strength of RO feed increased from 35,000 mg/L to 58,000 mg/L in the range of RO concentrate at recovery 40 %, the MFI-UF was observed to increase by 19 %.

The observed effect of ionic strength on seawater (RO feed) and DCW may depend on the particles (concentration and nature) present in the feed water. Faibish et al. (1998) reported that bigger particles required higher ionic strength solution to reduce repulsive forces between particles. A significant decrease in cake layer porosity with increasing ionic strength was observed for particles smaller than 47 nm while no significant changes was observed for particles bigger than 310 nm (Faibish et al., 1998).

Additionally, organic compounds can interfere with ionic strength effect on MFI-UF values. Ghosh and Schnitzer (1980) reported that NOM components at low ionic strength are flexible linear macromolecules while at high ionic strength they are rigid compounds. Flexible macromolecules can pass through a membrane easier than the rigid macromolecules. But it was reported that this is not the case for a charged nanofiltration membrane, since charged linear macromolecules may be retained by charge rejection (Braghetta et al., 1997). The effect of ionic strength on bovine serum albumin (BSA) was studied as well, and the results are presented in the annex 7.9.2.2. It was observed that the higher the ionic strength, the lower the MFI-UF value measured. The intermolecular adhesion forces among BSA molecules decrease with increasing ionic strength, resulting in higher cake porosity leading to a decrease in MFI-UF. This behaviour is attributed to the conformational changes of BSA molecules (Xu and Logan, 2005).

The MFI-UF of RO feed was observed to increase with increasing ionic strength. This can be attributed to compression of the double layer around the particles/ colloids and reduction of their inter-particles distance as a result of the increase in ionic strength.

The increase in salinity in canal water and in RO feed seawater produced higher MFI-UF values In the salinity range of seawater for a 40 % recovery, the increase for canal water was 60 and 10 %, for RO feed was about 19 %. In these two cases the particles concentration was kept constant as no dilution occurred.

### 7.5.1.3   RO concentrate dilution

In the case of the RO unit (site A) where the recovery is 45 %, the expected concentrate concentration will be ~1.82 times the feed concentration. This increase in ionic strength along the RO system, may affect the MFI measurements to calculate the deposition factor. In this case, the RO concentrate sample was diluted as shown in Figure 7.10 to a value close to the RO feed.

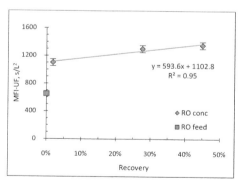

**Figure 7.10. MFI-UF values for RO concentrate dilutions at 250 L/m²-h and 30 kDa PES membranes - Site A**

The dilution of the RO concentrate showed a linear trend with MFI. In addition the corresponding MFI value for the RO feed is presented. Interestingly, the diluted RO conc MFI value at 0 % (~1100 s/L²) is higher than the RO feed ($\sim650$ s/L$^2$) even though the dilution also affects the particles/colloids concentration.

The percentage decrease in MFI-UF from 45 % to 2 % recovery for RO concentrate sample was $\sim22$ %.

In Figure 7.11 a second test is presented with water from site B. In this case the recovery of the RO is 40 %.

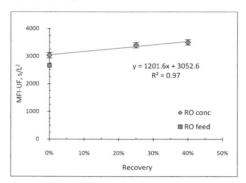

**Figure 7.11. MFI-UF values for RO concentrate dilutions at 250 L/m²-h and 10 kDa PES membranes - Site B**

From Figure 7.11, the percentage decrease in MFI-UF from 45 % to 0 % recovery for RO concentrate sample was $\sim15$ %.

When diluting the RO concentrate to a salinity level comparable with RO feed, the decrease in MFI values was around 22 and 15 %.

The MFI-UF constant flux test has a standard deviation about 10 %. In this sense the measured salinity effect is about 10 % significant. This may suggest that a correction factor could be introduced when comparing RO feed and RO concentrate MFI-UF values.

In previous sections was observed that the increase in ionic strength inside the pressure vessel influences the measured MFI values in the RO concentrate and therefore affects the deposition factor, suggesting that a correction factor for ionic strength (IScf) should be considered.

In Figure 7.12 it is illustrated, based on Eq. 7.17., the effect on deposition factor of correcting for ionic strength effect the measured RO concentrate MFI-UF values. In the left figure, for a recovery of 50 %, the corrected deposition factor is projected for various percentages of correction for ionic strength (i.e., 0, 5, 10, 15, and 20 %). It can be observed that the introduction of a correction factor increases the measured deposition factor; for instance, the deposition factor increases from 0 to 0.2 for 10 % IScf in deposition factor or from 0.25 to 0.33 for 5 % IScf.

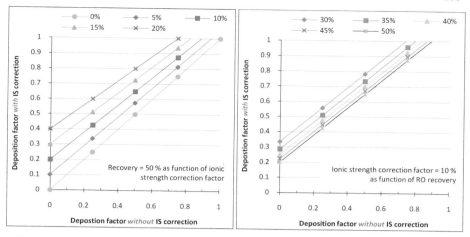

**Figure 7.12. Deposition factor with and without ionic strength correction as a function of: percentage ionic strength correction in RO concentrate's MFI value (left) and as function of recovery for 10 % IS correction (right)**

In the right figure it is illustrated the effect of 10 % IScf at various RO recoveries. For instance, considering a measured deposition factor equal to 0, the corrected deposition factor at 30 % recovery is 0.33 while at 50 % recovery is 0.2.

Correcting the measured MFI-UF values in the RO concentrate for ionic strength effect shows to be significant. This correction can be performed in two ways: i) diluting the RO concentrate water in a way that only salinity is affected, and ii) applying a percentage for IScf.

In the following sections, for location A and B, the deposition factor values are reported as they were measured without diluting the RO concentrate. A 10 % ionic strength correction factor was introduced for comparison.

## 7.5.2   LOCATION A

### 7.5.2.1   Measurements in parallel

The MFI-UF values were measured for the RO feed and RO concentrate, and the deposition factors calculated. The results are presented in Table 7.4 and Figure 7.13.

The deposition factor values were also measured on different dates. In all cases the values are more than 0 and less than one. The MFI-UF values for the RO concentrate are reported as they were measure and no correction factor due to ionic strength was applied.

**Table 7.4. MFI-UF values at 250 L/m²-h and deposition factor (Ω) in RO unit - Site A**

| Date | kDa | Mat. | R | RO feed | RO conc | Ω | Ω with 10 % IScf |
|---|---|---|---|---|---|---|---|
| 01.12.09 | 10 | PES | 45% | 780 | 1320 | 0.15 | 0.36 |
| 04.12.09 | 100 | PES | 45% | 230 | 240 | 0.95 | 1.07 |
| | 50 | PES | 45% | 790 | 1350 | 0.13 | 0.34 |
| | 30 | PES | 45% | 640 | 1150 | 0.03 | 0.25 |
| | 10 | PES | 45% | 1180 | 2050 | 0.10 | 0.31 |
| 05.12.09 | 100 | PES | 45% | 360 | 375 | 0.95 | 1.08 |

The MFI values increase as the MWCO decreases. At 100 kDa the measured values are almost the same (230 and 240), while for 10 kDa the difference between RO concentrate and RO feed is very clear (1.73 times). Unexpectedly, the obtained values for 50 kDa are slightly higher than for the 30 kDa for both cases of RO feed and RO concentrate.

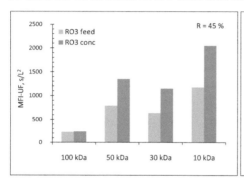 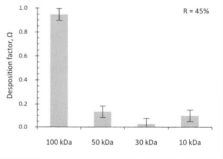

**Figure 7.13. MFI-UF values at 250 L/m²-h for RO feed and RO concentrate (left) and Deposition factor (right) as function of MWCO**

Figure 7.13 (right) shows the calculated deposition factors as a function of membrane MWCO. This figure suggests that particles and colloids larger than 100 kDa deposit on the membranes (0.95). In the case of 50, 30 and 10 kDa, the deposition factors are 0.13, 0.03 and 0.10, respectively.

### 7.5.2.2   Measurements in series

Table 7.5 and Figure 7.14 present the obtained results for the fractionation in series.

**Table 7.5. MFI values and deposition factor (Ω) in RO unit - Site A**

| Date | kDa | Range. | R | RO feed | RO conc | Ω | Ω with 10 % IScf |
|---|---|---|---|---|---|---|---|
| 04.12.09 | 100 | >100 | 45% | 360 | 375 | 0.95 | 1.08 |
| | 50 | 100-50 | 45% | 900 | 1450 | 0.25 | 0.45 |
| | 30 | 50-30 | 45% | 760 | 1320 | 0.10 | 0.31 |
| | 10 | 30-10 | 45% | 1140 | 1250 | 0.88 | 1.02 |

The MFI values in the RO feed increased by 150 % at 50 kDa, decreased by 16 % at 30 kDa and increased by 50 % at 10 kDa. The MFI values for RO concentrate increased by 287 % at 50 kDa, decreased by 9 % at 30 kDa and decreased by 5 % at 10 kDa.

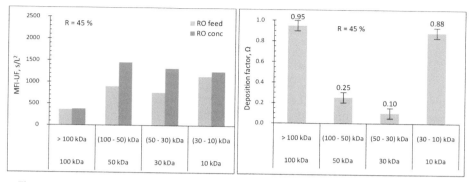

**Figure 7.14. MFI-UF values for RO feed and RO concentrate fractions (left) and deposition factor (right) as function of particle size range**

Similar to the previous section, the deposition factor with the 100 kDa membranes was close to 1. For the fractions 100-50 and 50-30 kDa, the deposition factors are 0.25 and 0.10, respectively, while for the fraction 30-10 kDa, the deposition factor is ~0.9.

These results suggest that particles bigger than 100 kDa and smaller than 30 kDa are most likely accumulating on the surface of the membranes. Conversely, in the case of particles smaller than 100 kDa and bigger than 30 kDa, the deposition factor values suggest these particles do not attach to the membranes. These results also suggest that there is a selective deposition of particles in the RO system.

Furthermore, it is possible to compare the MFI values measured in parallel with the partial sum of the MFI values obtained in series. The comparison of the summed values may suggest that particle size distribution plays a role in the deposition factor measurements. This is illustrated in Table 7.6.

**Table 7.6. Comparison of MFI values in parallel and in series for RO feed and RO concentrate**

| RO feed | | | | | RO concentrate | | | | |
|---|---|---|---|---|---|---|---|---|---|
| Parallel | | Series | | | Parallel | | Series | | |
| kDa | MFI | kDa | MFI | D* | kDa | MFI | kDa | MFI | D* |
| 100 | 230 | 100 | | | 100 | 240 | 100 | | |
| 50 | 790 | 50+100 | 1260 | 1.6 | 50 | 1350 | 50+100 | 1825 | 1.4 |
| 30 | 640 | 30+50+100 | 2020 | 3.2 | 30 | 1150 | 30+50+100 | 3145 | 2.7 |
| 10 | 1180 | 10+30+50+100 | 3160 | 2.7 | 10 | 2050 | 10+30+50+100 | 4395 | 2.1 |

*D = ratio of sum MFI values in series over MFI value in parallel

In all cases, the sum of the partial MFI values obtained in series yields a higher total than the MFI values measured in parallel. This may suggest that narrower particle size distribution in the in series tests yields more compact cakes and thus higher MFI values.

## 7.5.3   LOCATION B

### 7.5.3.1   Measurements in parallel

As for location A, in this section the MFI values are presented for RO feed and RO concentrate and the calculated deposition factor values.

In two cases negative $\Omega$ values were observed (-0.03 and -0.05). However these values could be considered as zero as the MFI test has 10 % standard deviation.

**Table 7.7. Measured MFI values and deposition factor ($\Omega$) in RO unit**

| Date | kDa | Mat. | R | RO feed | RO conc | $\Omega$ | $\Omega$ with 10 % IScf |
|---|---|---|---|---|---|---|---|
| 23.04.09 | 100 | RC | 40 % | 190 | 205 | 0.88 | 1.04 |
| 28.04.09 | 100 | RC | | 127 | 135 | 0.91 | 1.06 |
| 16.06.09 | 100 | PES | | 395 | 430 | 0.87 | 1.03 |
| 02.07.09 | 100 | RC | | 203 | 220 | 0.87 | 1.04 |
| 06.07.09 | 100 | PES | | 200 | 205 | 0.96 | 1.12 |
| 10.05.10 | 100 | PES | | 980 | 1650 | -0.03 | 0.23 |
| | 50 | PES | | 2350 | 3850 | 0.04 | 0.29 |
| | 10 | PES | | 5975 | 10170 | -0.05 | 0.20 |
| | 5* | PES | | 2900 | 4400 | 0.22 | 0.45 |

\* Sample measured at 15 L/m²-h. The other values were measured at 250 L/m²-h.

The MFI values increase as the MWCO decreases. When a 5 kDa membrane was tested, a 15 L/m²-h flux was used; in all the other cases the flux was 250 L/m²-h.

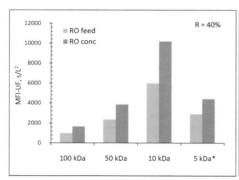 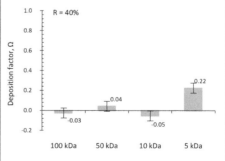

**Figure 7.15. MFI-UF values for RO feed and RO concentrate (left) and Deposition factor (right) as function of MWCO**

Figure 7.15 (right) presents the calculated deposition factors as a function of membrane MWCO. This figure suggests that particles bigger than 10 kDa do not accumulate on the membranes. The measured $\Omega$ value for 5 kDa was about 20 %.

For the four membranes tested, the measured deposition factors are close to 0. This indicates that particles are barely accumulating on the surface of the RO membranes.

### 7.5.3.2 Measurements in series

Table 7.8 and Figure 7.16 present the obtained results.

**Table 7.8. MFI values and deposition factor ($\Omega$) in RO unit - Site B**

| Date | kDa | Range. | R | RO feed | RO conc | $\Omega$ | $\Omega$ with 10 % IScf |
|---|---|---|---|---|---|---|---|
| 06.07.09 | 100 | >100 | 40% | 199 | 205 | 0.95 | 1.1 |
| | 50 | 100-50 | 40% | 1850 | 1940 | 0.93 | 1.08 |
| | 30 | 50-30 | 40% | 1205 | 1870 | 0.17 | 0.4 |
| | 10 | 30-10 | 40% | 16450 | 14400 | 1.19 | 1.32 |

The particle fraction 50-30 kDa has a significantly different deposition factor in comparison with the fractions bigger than 50 kDa and less than 30 kDa. This suggests that this 50-30 kDa fraction is not accumulating on the surface of the membranes while the other fraction most likely do.

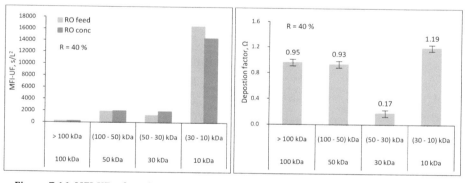

**Figure 7.16. MFI-UF values for RO feed and RO concentrate fractions (left) and deposition factor (right) as function of particle size range**

Similar to the previous section, the deposition factor with the 100 kDa membranes was close to 1. For the fractions 100-50 and 50-30 kDa, the deposition factors are 0.93 and 0.17, respectively, while for the fraction 30-10 kDa, the deposition factor is ~1.2.

## 7.6 Conclusions

From the previous sections, the following can be stated:

- The effect of salinity was studied for fresh water and for seawater. In general, an increase in MFI-UF values with increasing the salinity level in the solution was observed. In general, for a feedwater around 3.5‰ salinity and considering a RO recovery of around 40 %, the measured MFI-UF increase was ~15 % with respect to the MFI value in the RO feed.
- Correcting for ionic strength effect in the RO concentrate's MFI value increases the value of the deposition factor.
- Measured deposition factors varied between 0 and 1, depending on location and MFI pore size, which indicate differences in properties of the particles present.
- In many cases, a different MWCO used in the test in parallel produced different deposition factor values for the same water sample. This might be an indication that particle size distribution influences the results.

## 7.7    Acknowledgments

Special thanks to Eduard Gasia Bruch, Carmen Galvan and Xavier Serrallach from *CETAQUA* (Spain) for their logistic support during the tests and help with the data collection from the pilot plant. Thanks to Rinnert Schurer for the help in the tests in the North Sea plant.

## 7.8    List of abbreviations and symbols

### 7.8.1    ABBREVIATIONS

| | |
|---|---|
| DMF | Dual media filtration |
| kDa | Kilo Dalton |
| MFI-UF | Modified fouling index – ultra filtration |
| MWCO | Molecular weight cut off |
| PES | Polyethersulfone |
| RC | Regenerated cellulose |
| RO | Reverse osmosis |
| SWRO | Seawater reverse osmosis |
| UF | Ultra filtration |

### 7.8.2    SYMBOLS

| | |
|---|---|
| $A$ | Effective membrane surface area ($m^2$) |
| $C_b$ | Concentration of particles in a feed water ($kg/m^3$) |
| $d_p$ | Diameter of particles forming the cake (m) |
| $I$ | Fouling index of particles in water to form a layer with hydraulic resis. ($m^{-2}$) |
| $J$ | Permeate water flux ($m^3/m^2{\cdot}s$) |
| $R_c$ | Cake formation resistance ($m^{-1}$) |
| $R_m$ | Membrane resistance ($m^{-1}$) |
| $V$ | Filtrate volume ($m^3$) |
| $\alpha$ | (Average) specific cake resistance (m/kg) |

| $\Omega$ | Deposition factor (-) |
|---|---|
| $\varepsilon$ | Cake / membrane surface porosity (-) |
| $\eta_T$ | Water viscosity at temperature T (N·s/m$^2$) |
| $\rho_p$ | Density of particles forming the cake (kg/m$^3$) |
| $\tau$ | Tortuosity of membrane pores |
| $\omega$ | Compressibility coefficient (-) |
| $\psi$ | Cake ratio (-) |
| $e$ | elementary charge |
| $N_A$ | Avogadro's number |
| $IS$ | ionic strength |
| $IScf$ | ionic strength correction factor |
| $\varepsilon_r$ | dielectric permittivity of water |
| $\varepsilon_0$ | permittivity of free space |
| $k$ | Boltzmann constant |

## 7.9    Annex

### 7.9.1    SWRO DESIGN PROJECTIONS

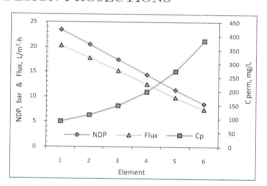

**Figure 7.17. NDP, flux and permeate concentration projections for a 15 m³/h SWRO system using SWC6 elements (Recovery = 40 %)**

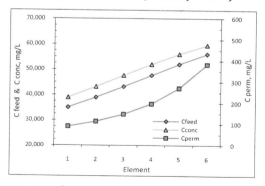

**Figure 7.18. Feed, permeate and concentrate concentration projections for a 15 m³/h SWRO system using SWC6 elements (Recovery = 40 %)**

## 7.9.2  SALINITY EFFECT

### 7.9.2.1  NaCl grade

Two grades of NaCl salt were compared (99.5 and 99.9999 %). The obtained results are shown in Figure 7.19.

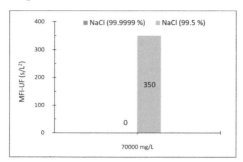

**Figure 7.19. MFI-UF for two grades of NaCl salt at 70,000 mg/L**

The NaCl 99.9999 % is free of particles as no fouling was observed while for NaCl 99.5 % fouling was observed.

### 7.9.2.2  Albumin + NaCl solution

Fouling potential under the influence of various solution chemistries is determined mostly by the intermolecular forces among foulants in the cake layer. Lee and Elimelech (2006) reported that there is a strong relationship between the flux decline and foulant-foulant adhesion forces. The influence of solution ionic strength on 5 mg/L bovine serum albumin (BSA) is presented in the figure below. The fouling tests were carried out at three NaCl concentrations: 0, 35, 70 g/L.

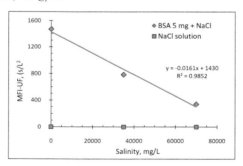

**Figure 7.20. MFI-UF of BSA 5 mg/L as function of salinity**

Figure 7.20 indicates that with increasing ionic strength fouling potential decreases; this is due to the intermolecular adhesion forces among BSA molecules decreasing with increasing ionic strength, resulting in higher cake porosity and leading to a decrease in MFI-UF. This behaviour is attributed to the conformational changes of BSA molecules (Xu and Logan, 2005).

# 7.10  References

AIKEN, G. & LEENHEER, J. 1993. Isolation and chemical characterization of dissolved and colloidal organic matter. *Chemistry and Ecology,* 8, 135-151.

BELFORT, G., DAVIS, R. H. & ZYDNEY, A. L. 1994. The behavior of suspensions and macromolecular solutions in crossflow microfiltration. *Journal of Membrane Science,* 96, 1-58.

BOERLAGE, S. F. E. 2001. *Scaling and Particulate Fouling in Membrane Filtration Systems,* Lisse, Swets&Zeitlinger Publishers.

BOERLAGE, S. F. E., KENNEDY, M. D., ANIYE, M. P. & SCHIPPERS, J. C. 2003. Applications of the MFI-UF to measure and predict particulate fouling in RO systems. *Journal of membrane science,* 220, 97-116.

BRAGHETTA, A., DIGIANO, F. A. & BALL, W. P. 1997. Nanofiltration of natural organic matter: pH and ionic strength effects. *Journal Environmental Engineering,* 123, 628-641.

CHELLAM, S. & WIESNER, M. R. 1998. Evaluation of crossflow filtration models based on shear-induced diffusion and particle adhesion: Complications induced. *Journal of Membrane Science,* 138, 83-97.

CRITTENDEN, J. C., TRUSSELL, R. R., HAND, D. W., HOWE, K. J. & TCHOBANOGLOUS, G. 2005. Membrane filtration. *In:* JOHN WILEY & SONS, I. (ed.) *Water Treatment: Principles and Design / MWH.* Second ed. New Jersey: Montgomery Watson Harza (Firm).

DAVIS, R. H. 1992. Modeling of fouling of crossflow microfiltration membranes. *Separation and purification methods,* 21, 75-126.

DAVIS, R. H. & SHERWOOD, J. D. 1990. A similarity solution  for steady-state cross flow microfiltration. *Chemical Engineering Science,* 45, 3203-3209.

FAIBISH, R. S., ELIMELECH, M. & COHEN, Y. 1998. Effect of Interparticle Electrostatic Double Layer Interactions on Permeate Flux Decline in Crossflow Membrane. *Journal of Colloid and Interface Science,* 204, 77-86.

GHOSH, K. & SCHNITZER, M. 1980. Macromolecular structures of humic substances. *Soil science,* 129, 266-276.

HOEK, E. M. V., KIM, A. S. & ELIMELECH, M. 2002. Influence of Crossflow Membrane Filter Geometry and Shear Rate on Colloidal Fouling in Reverse Osmosis and Nanofiltration Separations. *Environmental engineering science,* 19, 357-372.

IUPAC 1971. *Manual of symbols and terminology for physicochemical quantities and units, Appendix II definitions, Terminology and symbols in colloid and surface chemistry,* D. H. Everet.

JIANG, T. 2007. *Characterization and modelling of soluble microbial products in membrane bioreactors,* Ghent, Ghent University.

LEE, S. & ELIMELECH, M. 2006. Relating organic fouling of reverse osmosis membranes to intermolecular adhesion forces. *Environmental science & technology,* 40, 980-987.

MENDRET, J., GUIGUI, C., SCHMITZ, P. & CABASSUD, C. 2009. In situ dynamic characterisation of fouling under different pressure conditions during dead-end filtration. *Journal of Membrane Science,* 333, 20-29.

SCHIPPERS, J. C. 1989. *Vervuiling van hyperfiltratiemembranen en verstopping van infiltratieputten,* Rijswijk, Keuringinstituut voor waterleidingartikelen KIWA N.V.

SCHIPPERS, J. C., FOLMER, H. C. & KOSTENSE, A. Year. The effect of pre-treatment of river rhine water on fouling of spiral wound reverse osmosis membranes. *In:* 7th International Symposium on Fresh water from the Sea, 1980. 297-306.

SCHIPPERS, J. C., HANEMAAYER, J. H., SMOLDERS, C. A. & KOSTENSE, A. 1981. Predicting flux decline or reverse osmosis membranes. *Desalination,* 38, 339-348.

SIOUTOPOULOS, D. C., YIANTSIOS, S. G. & KARABELAS, A. J. 2010. Relation between fouling characteristics of RO and UF membranes in experiments with colloidal organic and inorganic species. *Journal of Membrane Science,* 350, 62-82.

SONG, L. F. & ELIMELECH, M. 1995. Particle deposition onto a permeable surface in laminar flow. *Journal of Colloid and Interface Science,* 173, 165-180.

STUMM, W. & MORGAN, J. J. 1996. *Aquatic chemistry: chemical equilibria and rates in natural waters,* New York, Wiley interscience publication.

XU, L. C. & LOGAN, B. E. 2005. Interaction forces between colloids and protein-coated surfaces measured using an atomic force microscope. *Environmental science & technology,* 39, 3592-3600.

# Chapter 8

## 8  The role of particles/colloids in SWRO systems - Fouling prediction and validation

Chapter 8 is based on:

SALINAS RODRÍGUEZ, S. G., KENNEDY, M. D., AMY, G. & SCHIPPERS, J. C. (2011). The role of particles/colloids in SWRO systems - Fouling prediction and validation. *Water Research*, submitted.

SALINAS RODRÍGUEZ, S. G., MAMOUN, A., SCHURER, R., KENNEDY, M. D., AMY, G. L. & SCHIPPERS, J. C. (2009). Modified fouling index (MFI-UF) at constant flux for seawater RO applications. In: EDS (ed.) *Desalination for the Environment: Clean water and Energy*. Baden-Baden, Germany: European desalination society.

## 8.1   Introduction

Reliable methods to predict the fouling potential of RO feedwater are important in preventing and diagnosing fouling at the design stage and for monitoring pre-treatment performance during plant operation (Boerlage, 2007). Particles are one of the possible causes of RO fouling. Several authors have also mentioned that several types of fouling may occur simultaneously (Khedr, 2000).

In water treatment systems using membrane technology it is a common practice to judge the quality of a feed/pre-treated water with help of fouling indices such as silt density index (SDI) and modified fouling index (MFI). These indices make use of a 0.45 μm filter and they are measured at constant pressure (~2 bar) that produce very high (initial) flux rates. SDI has a maximum possible value of 6.67 while the MFI values have no limit in measured values. RO membrane manufacturers have established an $SDI_{15} < 3$ as a recommended value to minimize particulate fouling (DOW, 2005).

As discussed in chapter 6, in cake formation there are several mechanisms affecting the measured resistance and therefore the fouling potential i.e., concentration of particles, flux effect on particles re-arrangement in the cake and cake compression during cake formation. Results presented in chapter 6 showed that at high flux rates (>80 L/m$^2$-h) the cake formation is highly affected by the rate of filtration and by the compression of the particles in the cake. This means that the cake properties (e.g., porosity, thickness) are significantly different when the cake is formed at low flux as in RO systems, compared to when the cake is formed at high flux rates as in the SDI or MFI tests. Flux rates in these tests can be as high as thousands of L/m$^2$-h depending on the water quality. This difference indicates that there is a big gap between the measured values by SDI and MFI and the possible particulate fouling occurring in RO systems.

Commonly before a RO membrane system, pre-treatment is applied. The purpose is to minimise fouling in any of its forms (particulate, organic, biological, or scaling). A proper pre-treatment is essential for the RO operation as it will increase the lifetime of the RO membrane and will maintain its performance (Fritzmann et al., 2007). In addition, pre-treatment prior to RO will reduce the frequency of chemical membrane cleaning.

The type of pre-treatment depends on the raw water quality and operational conditions. Conventional pre-treatment was mostly used in the past for SWRO systems. As the raw water quality declines, some alternatives for pre-treatment are being considered by using membrane filtration. Micro- and ultrafiltration membranes prior to SWRO are possible alternatives and it is estimated that their use will grow rapidly in the coming years (Fritzmann et al., 2007).

In addition, according to Busch et al (2010), UF stage costs have been
equivalent to, or in certain cases lower than, the conventional media filtration
process. Wolf et al. (2005) compared the pre-treatment between UF filtration
and conventional pre-treatment for SWRO which shows the typical cleaning
frequency for SWRO membrane is within the range of 1 – 12 months. More
details about the comparison are presented in Table 8.1.

**Table 8.1. A comparison of the impact of UF pre-treatment on an RO based sea water desalination plant**

| Characterization | UF pretreatment: ZeeWeed® 1000 immersed hollow fiber | Conventional pre-treatment: inline coagulation and 2 stage sand filters |
|---|---|---|
| Treated water quality | SDI < 2.5, 100% of the time, usually < 1.5 | SDI < 4 ~90% of the time |
| | Consistent, reliable quality | Fluctuating quality |
| | Positive barrier to particles and pathogens – no breakthrough. | Not a positive barrier to colloidal and suspended particles |
| | Turbidity: < 0.1 NTU | Turbidity: < 1.0 NTU |
| | Bacteria: > 5 log removal | |
| | Giardia : > 4 log removal | |
| | Virus: > 4 log removal | |
| Typical lifetime | UF Membranes: 5 - 10 years | Filter media: 20 - 30 years |
| | Cartridges : often not needed | Cartridges: 2 - 8 weeks |
| Average RO flux | ~ 18 L/m² h | ~ 14 L/m² h |
| SWRO replacement-rate | ~ 10 % per year | ~ 14 % per year |
| SWRO cleaning frequency | ~ 1 - 2 times per year | ~ 4 - 12 times per year |
| Pre-treatment foot-print | ~ 30 - 60 % (of conventional) | 100 % |

Source: Wolf et al. (2005)

Membrane cleaning is needed to restore performance by removing any fouling
layers which can cause a pressure increase or permeate flux decrease, and salt
rejection decrease. To avoid any further permanent membrane damage,
membrane cleaning must be carried out periodically (Fritzmann et al., 2007).
In addition to this, membrane cleaning should be performed when one or more
of the parameters mentioned below are achieved (DOW, 2005):

- The normalized permeate flow drops by ~10 %.
- The normalized salt passage increases by ~5-10 %.
- The normalized pressure drop (feed pressure minus concentrate pressure) increases by 10-15%.
- The net driving pressure increases by 10-15 %.

Cleaning frequency in RO plants depends on the performance of the system,
especially the pre-treatment efficiency.

To measure and accurately estimate particulate fouling in SWRO systems the
model must be as close as possible to the RO system. Thus, several aspects
must be considered as shown in Table 8.2.

**Table 8.2. Comparison between RO and fouling indices**

| Parameter | RO system | SDI, MFI | MFI-UF |
|---|---|---|---|
| Filtration mode | Cross flow | Dead-end | Dead-end |
| Operation | Constant flux | Constant pressure | Constant flux |
| Pore size | < 1 nm | 0.45 μm | 10 - 200 kDa (~2-30 nm) |
| Particle deposition | Due to shear force not all particles accumulate on the membrane surface. | Fouling indices accumulate all the particles bigger than filter pore size. | Particles deposition considered through Deposition Factor[1]. |
| Ionic strength | Increase in IS inside the pressure vessel | Not considered | IS factor considered in the Dep. Factor[2]. |
| Flux rate | Low (~10-20 L/m$^2$-h) | High (> 1000 L/m$^2$-h) | Low-High[3] |
| Criteria for membrane cleaning | Increase in NDP[4], ΔP[5], MTC[6] | Flux decline | Increase in NDP |
| Membrane material | Thin-film composite, | CA, PVDF, PAN, CN, | PES, RC |

[1]Chapter 7 discusses deposition of particles in RO systems. [2]Discussed in chapter 7. [3]MFI test usually is performed at 250 L/m$^2$-h for a quick test (30 min) compared with tests at 15 L/m$^2$-h that may last several hours. [4]Net driving pressure. [5]Feed-concentrate pressure. [6]MTC = Mass transfer coefficient.

The MFI models – constant pressure and constant flux – to predict particulate fouling were proposed initially by Schippers et al. (1989, 1981) and later on applied in fresh water applications by Boerlage et. al (2001, 2003), The models are based on the assumption that particulate fouling on the surface of reverse osmosis (or nanofiltration) membranes can be described by the cake filtration mechanism. Belfort et al. (1994, 1979) reported that cake filtration is the main mechanism in cross flow filtration and in particulate fouling in RO systems. These models are valid when scaling, pore blocking and biofouling do not contribute to the fouling observed. The model to predict particulate fouling in SWRO systems using the MFI test is affected by particle deposition factor, MWCO membrane used in MFI test, and flux effect from the difference of flux in MFI test and flux in the real SWRO plant.

Several researchers have made use of these models to estimate the rate of particulate fouling in RO, however, with some limitations. For instance, in some cases negative deposition factors were found or cake compression effects were ignored. Thus, the existing prediction model equation using MFI-UF constant flux to predict fouling potential (cleaning time) still needs to be validated. Firstly, a membrane pore size for the MFI-UF test should be selected. Secondly, the deposition factor needs to be considered in the fouling prediction and be measured accurately. Thirdly, the difference between the applied flux in the MFI test and the flux in the real RO plant might influence greatly the projected results.

In this chapter, three different SWRO plants in different locations are studied. The objective is to investigate the role of particles in the fouling of SWRO systems by means of the MFI-UF constant flux test discussed in chapter 5 and to use the results to estimate the cleaning frequency of the RO elements. Also, conventional and membrane pre-treatment are evaluated.

## 8.2    Background

### 8.2.1    CAKE FILTRATION PREDICTION MODEL CONSTANT FLUX

In constant flux filtration the feed pressure increases to keep the production of the plant constant.

The output of the prediction model is the time needed to reach a certain additional pressure increase to maintain the same operational flux when a certain water quality is fed to the RO system. Furthermore, the time can be used as an indication of membrane cleaning frequency caused by particulate fouling.

Taking the standard equation describing the flux through a membrane:

$$\frac{dV}{A \cdot dt} = \frac{\Delta P}{\eta \cdot (R_m + R_c)}$$

Eq.    8.1

and substituting $J$ for flux and $R_c$ by:

$$R_c = \frac{V}{A} \cdot I$$

Eq.    8.2

the following equation is obtained:

$$J = \frac{\Delta P}{\eta \cdot \left(R_m + \frac{V}{A} \cdot I\right)}$$

Eq.    8.3

rewriting $V/A$ as $J \cdot t$ and rearranging gives (Boerlage et al., 2004):

$$\Delta P = J \cdot \eta \cdot R_m + J^2 \cdot \eta \cdot I \cdot t$$

Eq.    8.4

The fouling index $I$ can then be determined from the slope of the linear region of a plot of $\Delta P$ versus *time* which corresponds to cake filtration or by manipulation of Eq. 8.4. The MFI can be calculated using $I$ (from Eq. 8.4) for standard reference conditions as follows (Boerlage et al., 2004):

$$MFI = \frac{\eta_{20°C} \cdot I}{2 \cdot \Delta P_0 \cdot A_0^2}$$

Eq.    8.5

The MFI models – constant pressure and constant flux – to predict fouling developed by Schippers et al. (1989, 1981) are based on the assumption that particulate fouling on the surface of reverse osmosis (or nanofiltration) membranes can be described by the cake filtration mechanism. The relationship between the MFI measured for RO feed water and the flux decline predicted for a reverse osmosis system are derived below. When scaling, pore blocking and biofouling do not contribute to the fouling observed, the flux through a RO membrane is (Boerlage et al., 2004):

$$J_r = \frac{dV_r}{A_r \cdot dt_r} \qquad \text{Eq.} \quad 8.6$$

and,

$$J_r = \frac{\Delta P_r}{\eta_r \cdot (R_{mr} + R_{cr})} \qquad \text{Eq.} \quad 8.7$$

Where the subscript $r$ indicates that the parameter refers to filtration through a RO membrane. For a RO system operating under constant flux filtration, the time required for an increase in net driving pressure $NDP_r$ to occur can be predicted by manipulation of Eq. 8.4 applied to a RO membrane:

$$t_r = \frac{(NDP_r - NDP_{0r})}{J^2 \cdot \eta \cdot I_r} \qquad \text{Eq.} \quad 8.8$$

The relationship between $I_r$ and $I$ (from the MFI-UF measurement) is assumed as (Boerlage, 2001, Schippers, 1989):

$$I_r = \psi \cdot \Omega \cdot I \qquad \text{Eq.} \quad 8.9$$

where the cake ratio factor ($\psi$) accounts for differences between the cake deposited on the MFI membrane and that deposited on the RO membrane due to differences in flow rates, and the particle deposition factor ($\Omega$) represents the ratio of the particles deposited on the RO membrane to that present in the feedwater. The particle deposition factor is calculated from the relation between the $MFI$ of the concentrate at recovery $R$ of the RO system and the $MFI$ of the feedwater as follows:

$$\Omega = \frac{1}{R} + \frac{MFI_{concentrate}}{MFI_{feed}} \cdot \left(1 - \frac{1}{R}\right) \qquad \text{Eq.} \quad 8.10$$

The cake ratio factor ($\psi$) is taken into consideration by correcting the measured fouling index at a certain flux to a MFI value that would be obtained at similar RO operation, i.e., 15 L/m²-h. This correction can be achieved by measuring MFI-UF values at different filtration rates, i.e., 50, 150, 250, 350 L/m²-h and then extrapolating the obtained relation to the actual flux in a RO installation (i.e., 15 L/m²-h).

Assuming $I_r$ is related to $I$ from the MFI measurement (Eq. 8.5) and substituting in Eq. 8.9 with rearrangement, $I_r$ can be defined as:

$$I_r = \frac{2 \cdot MFI \cdot \psi \cdot \Omega \cdot \Delta P_0 \cdot A_0^2}{\eta_{20°C}} \qquad \text{Eq.} \quad 8.11$$

Combining Eq. 8.8 and Eq. 8.11 gives the time $t_r$ period in which the pressure of a RO system has increased from an initial operating pressure of $NDP_{or}$ to $NDP_r$ (Boerlage, 2001, Boerlage et al., 2004):

$$t_r = \frac{\eta_{20°C} \cdot (NDP_r - NDP_{0r})}{\eta_r \cdot J_0^2 \cdot \psi \cdot \Omega \cdot MFI \cdot 2 \cdot \Delta P_0 \cdot A_0^2} \qquad\qquad \text{Eq.} \quad 8.12$$

The parameters that are considered in the prediction model are summarized in the Figure 8.1. All of the mentioned parameters aim to represent RO particulate fouling. The outcome of the model is to obtain a time frame which means the estimated time to perform cleaning of the RO membranes when only particles are the cause of fouling (increase in pressure).

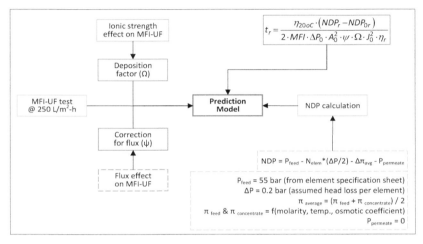

**Figure 8.1. Prediction model parameters**

The particle deposition factor can be measured on-site based on the MFI values of RO feed water and RO concentrate. On the other hand, it is known that the smaller the MWCO membrane that is used, the higher the MFI value that is obtained due to a higher retention of particles. As a consequence, a shorter predicted time for fouling will result from the model. All waters have different (site specific) characteristics, for instance the particle size distribution in the water may influence the fouling in the RO membranes, therefore, the same membrane MWCO may not fit all cases and this may need to be evaluated experimentally.

Regarding the flux effect, previous research showed that the higher the flux, the higher the MFI value. As a consequence, the shorter the predicted fouling time that is obtained from the model. As suggested by Boerlage (2001) and studied in chapter 6, this is caused by flux effect on cake porosity and cake compression at high flux rates. Meanwhile, the MFI test is based on cake filtration with no compression, which implies that the compression effect, as the cake formation occurs, is not incorporated.

Membrane surface properties such as roughness and charge have been reported to influence particulate fouling (Hoek et al., 2001). These possible effects are not considered on the model.

Hoek and Elimelech (2003) studied the fouling due to inorganic particles in RO membranes in a SEPA CF unit and reported that the particulate fouling mechanisms in RO are influenced by the increase in ionic strength due to rejection of ions present in water. They proposed a new model to include the ionic strength namely, cake enhanced osmotic pressure (CEOP). According to this model, "a thin foulant deposit layer may cause significant flux decline through enhanced salt concentration polarization". Chong et al. (2007, 2008) also reported that the effect of CEOP could be as much as the hydraulic resistance due to the foulant layer after studying colloidal silica and alginic acid.

Yiantsios et al. (2005) discussed that the difference in UF membranes not considering the extra resistance due to CEOP does not necessarily imply that a correlation of the fouling behaviour of different membranes should not exist.

More recently, Kovalsky et al. (2009, 2008) presented a mathematical model to predict the rate at which pressure increases in a dead-end system. This new model considers simultaneous cake consolidation and cake growth following the cake-layer approach discussed by several researchers (Rietema, 1953, Ruth, 1935, Tien and Bai, 2003, Tiller, 1958, Tiller and Kwon, 1998, Yim et al., 2003). This model has been tested with yeast solutions and depends on cake properties and shear as input parameters which makes it difficult to be applied.

## 8.3   Goal and objectives

The goal of this chapter is to study the role of particles and specifically the role of colloids in fouling of SWRO systems, especially regarding the particle size, cake deposition factor ($\Omega$), and cake factor ratio ($\psi$) which influence the particulate fouling rate of a SWRO membrane.

### 8.3.1   OBJECTIVES

The objective of this study focuses on applying the MFI-UF constant flux test and fouling prediction model to estimate the rate of particulate fouling in seawater reverse osmosis systems. For this:

MFI-UF tests were performed in three different locations with different pre-treatments. With the measured values the cleaning frequency of RO membranes was estimated.

## 8.3.2   RESEARCH QUESTION

How significant is the fouling potential of seawater in reverse osmosis systems?

# 8.4   Material and methods

## 8.4.1   ON SITE TESTING LOCATIONS

Three locations were selected for the tests. The locations and pre-treatment systems are summarized on Table 8.3.

**Table 8.3. Summary of the tested locations/plants**

| Location | Intake | Pre-treatment | RO unit |
|---|---|---|---|
| A (North-Western Mediterranean water) | Submerged pump (L = 2.5 km) | UF (0.02 μm) | R = 45 %. J = 15 L/m²-h. 7 elements/vessel. 8" mod. Never cleaned. |
| B (North Sea water) | Submerged pipe (L = 100 m) | Strainer – UF (~300 kDa) | R = 40 %. J = 15 L/m²-h. 6 elements/vessel. 8" mod. Never cleaned. |
| C (Northern Mediterranean water) | Direct intake | UF (0.01 μm) | R = 40 %, J = 15 L/m²-h. |

At site A, is possible to find three pre-treatment lines in parallel including dissolved air flotation (DAF) and dual media filtration (DMF); however, only UF was connected to a RO unit. A summary of the water qualities is presented on Table 8.4.

**Table 8.4. Summary of water characteristics**

| Sample | Location | DOC, mg/L | pH | T, °C | SUVA (L/mg·m) | EC, mS/cm | NTU |
|---|---|---|---|---|---|---|---|
| Raw water | A | 0.75 | 8.1 | 16.4 | 0.55 | 57-58 | 0.5-1.5 |
| UF perm / RO feed | A | 0.72 | 8.0 | 16.4 | 0.45 | 57-58 | |
| RO conc | A | 1.5 | | | 0.5 | | |
| Raw water | B | 1.45 | 8.1 | 13.5 | 2.3 | 48.5 | 8-12 |
| UF perm / RO feed | B | 1.3 | 6.5 | 13.5 | 2.15 | 48.5 | |
| RO conc | B | 2.2 | | | 2.0 | 76.5 | |
| Raw water | C | 1.15 | 8.21 | 18 | 0.75 | 57.1 | |
| UF perm | C | 0.85 | | | 0.7 | | |
| Coag + DMF effl. | C | 0.78 | | | 0.6 | | |

To reduce the impact of variations in water quality, as many tests as possible were performed on the same day.

## 8.4.2   SITE A DESCRIPTION

This site located 15 km from Barcelona, consists of two separate/independent treatment lines (W and T).

The W pilot plant consists of ultrafiltration followed by reverse osmosis at 51 % recovery, and the T pilot plant consists of coagulation + dissolved air flotation followed by ultrafiltration and then reverse osmosis at 45 % recovery. A second treatment line exists with double stage dual media filtration that at that time was not connected to a RO unit. The pilot plants receives water from an open intake (submerged pipe) located 2.5 km from the coast and 25 m below the surface of the water. This pipe is cleaned by chlorination (frequency not disclosed).

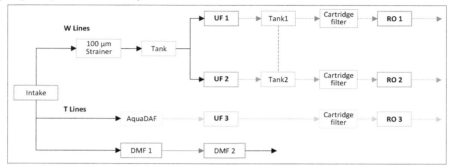

**Figure 8.2. Scheme of the Site A**

The average suspended solids of the raw water was about 3.5 mg/L and the average SDI value was 5.

## W lines

Before the raw water was fed to the UF, it passes through an 100 μm strainer. The ultrafiltration units (UF1 and UF2, identical) operate at constant flux at ~60 L/m²-h. Backwash is applied at double the operation flow with air scour every 30 min consisting of 10 seconds air scour, 15 seconds backwash with UF permeate and 45 seconds forward flush with raw water.

**Table 8.5. UF1 and UF2 units' description**

| Parameter | Value | Comment |
|---|---|---|
| Operation | Constant flux (1.9 m³/hr) | (typical pressure 0.7 bar) |
| Flux | ~58 L/m².h | |
| Nominal pore size | 0.03 μm | |
| Material | PVDF | |
| Brand | SFP-2660 | OM Exell - DOW |
| Backwash | 1.25 min | With air scour, permeate water |
| Chemically enhanced backwash (CEB) | Every 24 hours | ~ 48 cycles |
| Membrane area | 33 m² | |
| Filtration | Outside to inside | |

The RO systems consists of two units working in parallel (RO1 and RO2). Each unit has 6, 4" SW30-4040HR elements and operate at 52 % recovery. The RO2 unit has a hybrid configuration (at the beginning of the pressure vessel are placed high rejection modules and at the end are placed high production modules) while RO1 has a standard configuration. Both RO units operate at constant pressure (70 bar). Bisulphite and antiscalant are added at the front of the RO (Permatreat, phosphonated). Since operation of the plant (April 2009), the RO membranes were not cleaned. The RO production capacity is around 0.76 m$^3$/hr.

**T Lines**

An AquaDAF unit was not necessary during the testing period due to good raw seawater quality. The UF3 modules are Zenon membranes that operate in a submerged mode (Table 8.6).

**Table 8.6. UF3 unit description**

| Parameter | Value | Comment |
|---|---|---|
| Operation | | (typical pressure -1 to -15 psi) |
| Flux | 40-60 L/m$^2$-h | Estimated |
| Nominal pore size | 0.02 μm | |
| Material | PVDF | |
| Brand | ZeeWeed | Zenon |
| Membrane area | 55 m$^2$ | |
| Filtration | Outside to inside | |

There are two dual media filters (DMF) that operate in series. In DMF1 there are two layers of sand and pumice and in DMF2, sand and anthracite. DMF1 was not in operation during the testing period due to good raw seawater quality. Coagulation is used in combination with DMF. During the testing period the dose was 1.5 mg/L FeCl$_3$. The RO3 unit is one pressure vessel containing 7 8" elements, operated at 45 % recovery.

## 8.4.3    SITE B DESCRIPTION

The pilot plant is located in Noord-Beveland in The Netherlands. The plant (Figure 8.3) makes use of coagulation and ultrafiltration as a pre-treatment to the reverse osmosis units.

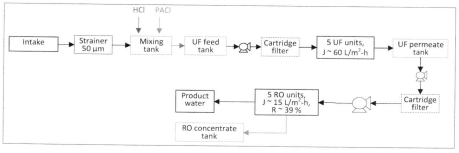

**Figure 8.3. Scheme of Site B**

Before coagulation, the pH is reduced to 6.5 and then coagulant (poly aluminium chloride, PACl) is added (0.5 mg/L as Al$^{3+}$) and mixing occurs mechanically and hydraulically in the mixing tank. Average conditions for the UF operation are described in Table 8.7.

**Table 8.7. UF unit description**

| Parameter | Value | Comment |
|---|---|---|
| Operation | Constant pressure | It can work at constant flux for short periods |
| Flux | ~60 L/m$^2$-h | After cleaning |
| Nominal pore size | ~300 kDa | |
| Material | PES | |
| Brand | SeaGuard | NORIT filtration |
| Membrane area | 37 m$^2$ | Per module |
| Filtration | Inside to outside | |
| Backwash | Every 45 min | |
| Cleaning | 1-2 x /day | |

The operational conditions in the plant such as coagulant dose, coagulant type and pH correction, are under experimental investigation and thus they changed over time.

## 8.4.4   SITE C DESCRIPTION

This site is located along the coast in the northern part of the Mediterranean sea. The pilot plant consists of two parallel reverse osmosis pre-treatment lines, namely: ultrafiltration (UF) and coagulation combined with dual media filtration (Coag+DMF).

The water intake is a submerged (2 m) pump close to the coast line (~5 m distance). Water is pumped (12 m height) for ~40 m length into a plastic tank for sand particles to settle and for algae removal through a 2 mm screen. Water is pumped again ~30 m in distance to the pilot plant installation. Seawater is received in a tank (0.8 m$^3$) and then distributed to the DMF line and to the UF line. There is no pH correction or acid addition in any part of the plant.

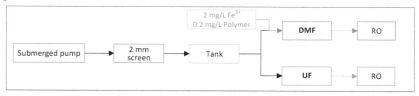

**Figure 8.4. Scheme of Site C**

The RO units consist of one 4" module working at 20 % recovery. For the prediction model estimation, typical full scale conditions were applied.

Iron chloride, FeCl$_3$ (40 % concentration), at a dose of 2 mg Fe$^{3+}$/L was employed. The residual iron in the water after DMF is around 10-20 μg/L. Coagulant is dosed in front of the DMF by dosing pumps.

A cationic polymer is added as well before DMF. The commercial polymer is Floerger FL 4520 – SEP, and the dose is around 0.15-0.2 mg/L. FL 4520 is a medium molecular weight, homopolymer of diallyldimethylammonium chloride (DADMAC). It is an effective organic coagulant for water and wastewater clarification in a wide variety of industrial, municipal and mining applications (SNF, 2009).

The DMF consists of a cylinder (~40 cm diameter) containing two media layers (anthracite and sand). Table 8.8 summarizes the characteristics of the unit.

**Table 8.8. DMF unit description**

| Parameter | Value | Comment |
|---|---|---|
| Flow rate | ~0.9 m³/h | Flow from the top to the bottom |
| Media | Anthracite (80 cm) | Anthracite is the upper layer |
| | Sand (80 cm) | |
| Anthracite | $d_p$ = 1.4 – 2.5 mm | $d_p$ = particle size |
| Sand | $d_p$ = 0.6 mm | |
| Cleaning | $\Delta P$ = 400 mbar | Backwash with air and water. It takes 1 hour |
| | | before new water is produced. |
| Total height of the DMF | 3.5 m | |
| Maturation period | 10 hours | According to the operator based on $SDI_{15}$ values. |
| $SDI_{15}$ | ~3.5 | Typical value after 10 hours. Includes coagulation. |

The pH of the water after Coag+DMF decreases from the raw water value (~8.21) to 7.8-8.0. The residence time from the raw seawater tank inside the installation to the DMF effluent sampling point is around 25 minutes.

The pilot plant has one UF module working permanently at a constant flux. The description of the UF is presented in Table 8.9.

**Table 8.9. UF unit description**

| Parameter | Value | Comment |
|---|---|---|
| Operation | Constant flux (1.6 m³/h) | (typical pressure ~70 kPa) |
| Flux | 57 L/m²-h | |
| Nominal pore size | 0.01 µm | Loose UF |
| Material | PVDF | |
| Brand | MEMCOR CMF-S S10V | Siemens |
| Backwash | 2 min | Air scour and backwash with |
| | | permeate water |
| Chemically enhanced backwash (CEB) | Every 40 cycles | Every ~ 10 hours |
| Membrane area | 27.9 m² | |
| Filtration | Outside to inside | |

The residence time from the raw seawater tank inside the pilot plant to the UF permeate sampling point is around 3 minutes.

## 8.4.5  OTHER CONSIDERATIONS

In the three plants and wherever possible the residence time of the water was considered to compare the performance of the various units. Also, to reduce the impact of variations in water quality, as many tests as possible were performed on the same day.

# 8.5  Results

## 8.5.1  PRE-TREATMENT ASSESSMENT

### 8.5.1.1  Site A – Ultrafiltration and Dual media filtration

From both pilot plants the MFI-UF values were measured at constant flux with a 10 kDa PES membrane. Initially for the raw water the MFI-UF was measured with a 30 kDa membrane. The obtained value was 1050 s/L$^2$. As this value was observed to be low in comparison with other seawaters, i.e., ~4500 s/L$^2$ with a 100 kDa for North Sea water, it was decided to perform the profiling of the pilot plants with a 10 kDa membrane. The obtained results are presented in Figure 8.5.

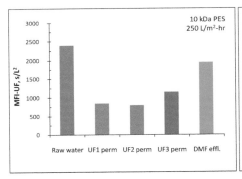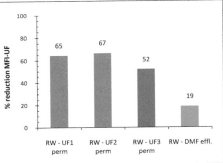

**Figure 8.5. MFI-UF values (left) and MFI-UF percentage reduction (right) for W and T lines**

The MFI-UF values for UF1 and UF2 are close 850 and 800 s/L$^2$, respectively. The MFI-UF value for UF3 was 1150 s/L$^2$. According the manufacturers the UF1 and UF2 have a nominal pore size of 0.03 μm and UF3 a nominal pore size of 0.02 μm. The lower MFI values in UF1/UF2 might be related to a narrower membrane pore size distribution in comparison with UF3 or due to integrity problems in UF3.

The DMF2 value was high at around 1,950 s/L$^2$. Unfortunately the operational data was not disclosed. Iron chloride was added at 1 mg/L as FeCl$_3$ in front of the DMF2.

The percentage reduction of MFI-UF values (Figure 8.5 right) before and after pre-treatment was 65 %, 67 %, 68 % and 19 % for UF1, UF2, UF3 and DMF2, respectively.

Unfortunately, the RO1 and RO2 units had an operational problem and did not work during the testing period. Therefore, the results from the following sections correspond to the RO3 unit.

## 8.5.1.2 Site B – Coagulation + Ultrafiltration

The MFI-UF plant profiling with a 100 kDa membrane considering several dates over the period 2009-2010 is presented in Table 5.7. The MFI values were measured at 250 L/m²-h. The samples on 10.05.10 are significantly higher than the ones in the previous year. This significant increase was correlated with an algae bloom and thus an increase in biopolymers (TEP) concentration in the raw water during the testing period.

**Table 8.10. MFI-UF (100 kDa) values in s/L² and percentage removal**

| Date | Raw water | UF feed | UF perm | Removal |
|------|-----------|---------|---------|---------|
| 23.04.09 | 4,310 | 2,935 | 190 | 94 % |
| 28.04.09 | 4,840 | 4,295 | 125 | 97 % |
| 16.06.09 | 3,800 | 3,650 | 395 | 89 % |
| 02.07.09 | 2,950 | 2,285 | 203 | 91 % |
| 06.07.09 | 2,840 | 2,450 | 200 | 92 % |
| 10.05.10 | 25,340 | 17,190 | 980 | 94 % |

Although the raw water values varied in time, on all the testing dates the MFI decrease after the UF was between 89 and 97 %.

Figure 8.6 shows the MFI-UF values measured with 100, 50 and 10 kDa membranes at 250 L/m²-h along a SWRO plant treating water from the North Sea for the higher foulant period. The plant is located in The Netherlands.

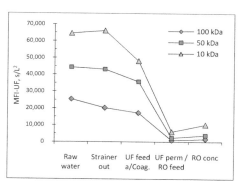

**Figure 8.6. MFI-UF plant profiling measured with 100, 50 and 10 kDa at 250 L/m²-h**

The percentages in reduction of MFI values after water passing through the ultrafiltration units were 94.3 %, 93.4 % and 87.6 % for 100, 50 and 10 kDa, respectively.

### 8.5.1.3    Site C – Coagulation + DMF and Ultrafiltration

As mentioned in the plant description, the pilot plant has two pre-treatment systems in parallel that treat the same raw water. For comparing both pre-treatments the MFI-UF values were measured before and after the units and with various membranes (100, 30 and 10 kDa) at 250 L/m²-h. The results are presented in Figure 8.7.

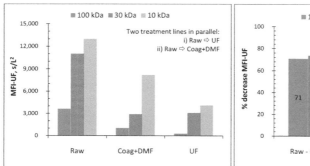 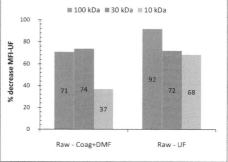

**Figure 8.7. MFI-UF values (left) and percentage removal (right) for raw seawater, Coag+DMF effluent and UF permeate with 100, 30 and 10 kDa membranes at 250 L/m²-h**

The raw seawater has an electrical conductivity equal to 57.1 mS/cm, a pH of 8.21 and, during the testing period the temperature in the water was 19° C.

The MFI-UF value for the raw water is comparable to North Sea water. The higher value is obtained with the smaller MWCO (10kDa) as more particles/colloids are retained by the membrane. The MFI-UF value for 30 kDa is between 100 and 10 kDa and closer to the latter's result. In general the same trend is observed for DMF and UF water, the values for the UF permeate being lower than for the DMF effluent.

The percentages in MFI-UF removal for each pre-treatment line are presented in Figure 8.7 right. For all of the MWCOs, the water passage through UF decreases MFI-UF values more in comparison with Coag+DMF. For UF, the MFI-UF decrease for 100, 30 and 10 kDa are 92 %, 72 % and 68 %, respectively. For Coag+DMF the reductions in MFI-UF values were 71 %, 74 % and 37 % for 100, 30 and 10 kDa, respectively.

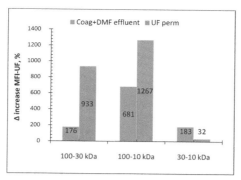

**Figure 8.8. Percentage additional increase in MFI-UF values between 100-30, 100-10 and 30-10 kDa membranes for raw seawater and Coag+DMF effluent and raw water and UF permeate with 100, 30 and 10 kDa membranes at 250 L/m²-h**

Figure 8.8 shows the additional increase in MFI-UF values by changing the MWCO (decreasing) for both pre-treatment lines. For the UF, particles with size between 100 and 30 kDa have a more significant effect on the MFI values than the ones between 30 and 10 kDa. For Coag+DMF, particles between 100 and 30 kDa and particles between 30 and 10 kDa showed a similar increase.

### 8.5.1.3.1   DMF profiling

The performance of the DMF was subject of testing. The tests started after backwashing of the unit to measure the evolution or change in effluent quality if any. Typically, the DMF cycle duration is 22 hours which corresponds approximately to 400 mbar head loss. For logistic issues, MFI-UF values were measured for the first 10 hours of the cycle. In parallel, $SDI_{15}$ and MFI 0.45 μm values were obtained with an automatic unit (MABAT SDI 2200) around every hour. The SDI and MFI 0.45 μm values were obtained with a cellulose acetate, 47 mm diameter membrane. The MFI-UF values were measured with a 30 kDa PES membrane. The results are presented in Figure 8.9.

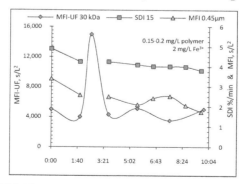

**Figure 8.9. MFI-UF, MFI 0.45μm and $SDI_{15}$ values for Coag+DMF effluent for the first 10 hours of DMF operation**

The SDI and MFI 0.45 μm results are of the same order of magnitude ($<5$). SDI values are higher than MFI 0.45 μm values. After ten hours of operation,

the SDI values decreased by 22 %, the MFI 0.45 μm decreased by 48 % and
no reduction was observed in MFI-UF (30 kDa).

From the monitoring system of the pilot plant, the particle count, head loss
pressure and turbidity in the DMF unit were obtained. These values are
presented in Figure 8.10. After ten hours of operation the head loss pressure
was 120 mbar. The turbidity values are not constant, with an average value
0.045 NTU.

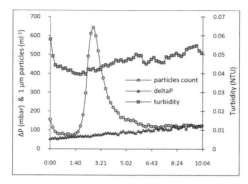

**Figure 8.10. DMF profiles for ΔP, 1 μm particles count and turbidity during the MFI-UF testing**

As can be observed above, a similar trend was observed between the MFI-UF
values (in Figure 8.9) and the particles count (in Figure 8.10), in particular
between hours 2 and 4. At the second hour of DMF operation, the polymer
dosing pump stopped working for almost 2 hours. Unfortunately, during this
period the SDI automatic unit did not record any value. However, a sample
was taken for MFI-UF.

## 8.5.2   PREDICTION MODEL

There are several parameters influencing the fouling prediction model, namely:
flux effect, membrane MWCO, and particle deposition factor. A summary of
the parameters used in the prediction model are described in Table 8.11.

**Table 8.11. Information for the prediction model in the various locations**

| Parameter | Site A | Site B | Site C | Comment |
|---|---|---|---|---|
| Temperature, °C | 16.4 | 10.5 | 19 | During testing period |
| TDS water, mg/L | 39,390 | 35,030 | 39,390 | |
| Recovery | 45 % for RO3, 52 % for RO1 & RO2. | 40 % | 40 % | During testing period |
| Feed pressure RO | 55 bar | 58.5 bar | 55 bar | |
| $NDP_0$ | 15.6 bar | 28.2 bar | 17.5 bar | Calculated for Site A water |
| $(NDP_r - NDP_{0r})$ | 2.33 bar | 4.23 bar | 2.62 bar | Corresponds to 15 % change |
| Deposition factor, Ω | Measured on-site. In some cases the worst case scenario is considered ($\Omega=1$, all particles are accumulated) | | | |
| Cake ratio factor, ψ | Considered based on MFI vs. flux relations | | | |

The particle deposition in RO systems was discussed in chapter 7 and flux effect on cake formation was discussed in chapter 6.

### 8.5.2.1    Site A

### 8.5.2.1.1    Flux effect

The MFI-UF values at various flux rates for a 30 and 10 kDa membranes are presented in Figure 8.11 for raw water and for UF3 permeate. In both cases, a linear trend was observed between MFI and flux.

From the measured trends it is possible to extrapolate, assuming a linear relation, the corresponding MFI values for a flux similar to RO in full scale operation (~15 L/m$^2$-h). In the case of raw water, for both 30 and 10 kDa membranes the projection gives a negative MFI value (-324 s/L$^2$ for 30 kDa and -605 s/L$^2$ for 10 kDa) meaning that particulate fouling potential is minimum at flux rates as low as 15 L/m$^2$-h.

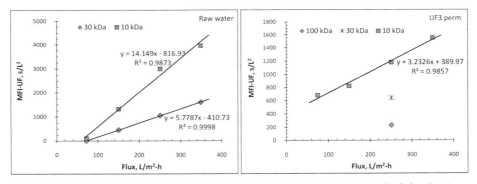

Figure 8.11. MFI-UF values vs. Flux for raw seawater (left) and UF3 perm (right) – Site A

The MFI-UF values with a 10 kDa PES membrane were measured for UF3 permeate at different fluxes (Figure 8.11). As UF3 has a nominal pore size of 0.02 μm, the particles in UF3 are smaller than 0.02 μm and have a narrower size distribution in comparison with raw water. Additional to the 10 kDa results, in Figure 8.11 are included the MFI-UF values for 30 and 100 kDa (640 and 230 s/L$^2$, respectively). In this case, the projected MFI-UF value for 15 L/m$^2$-h flux is ~435 s/L$^2$.

In comparison with the observed effect for the raw water in the previous section, for the UF3 permeate at 15 L/m$^2$-h the MFI-UF value is positive and around 435 s/L$^2$. This may be attributed to the narrower particle size distribution present in UF permeate that may create a less porous cake and therefore higher specific cake resistance; or particles creating porous cake were partly removed by UF unit.

### 8.5.2.1.2    Particles and colloids deposition in RO

The deposition factor values were measured on site. The results are presented in Table 8.12.

**Table 8.12. Deposition factor ($\Omega$) values at various MWCO in RO3 unit - Site A**

| MWCO, kDa | R | RO feed | RO conc | $\Omega$ | $\Omega$ with 10 % IScf |
|-----------|-----|---------|---------|------|--------------------------|
| 100 | 45% | 230 | 240 | 0.95 | 1.07 (1) |
| 50 | 45% | 790 | 1,350 | 0.13 | 0.34 |
| 30 | 45% | 640 | 1,150 | 0.03 | 0.25 |
| 10 | 45% | 1,180 | 2,050 | 0.10 | 0.31 |

### 8.5.2.1.3    Cleaning frequency

In Table 8.11 is summarized the data used to estimate the RO cleaning time at site A. Membrane cleaning is commonly practiced when a 15-20 % decrease in the normalised flux or increase in net driving pressure (NDP, in the front element) of an installation is observed. The measured MFI-UF values obtained for the raw water, DMF2 effluent and UF3 permeate were used to predict the time for a 15 % increase in NDP.

The results of the prediction model for UF3 permeate, DMF2 effluent and for raw seawater are presented in Table 8.13. The projections were performed considering a RO unit with recovery = 45 % and flux = 15 L/m²-h. Also presented in the same table are the "$t_r$" results with and without correction for flux (250 and 15 L/m²-h).

**Table 8.13. Estimated times for 15 % increase in NDP for raw seawater, UF permeate and Coag+DMF effluent - Site A**

| Sample | MWCO | $\Omega$ | MFI @ 250 lmh*, s/L² | $t_r$ @ 250 lmh, months | MFI @ 15 lmh, s/L² | $t_r$ @ 15 lmh, months | 1 bar, $t_r$ @ 15 lmh, months |
|--------|------|------|------|------|------|------|------|
| UF1 perm | 10 kDa | $1^1$ | 850 | 5.4 | 94 | 49 | 28.2 |
| UF2 perm | 10 kDa | $1^1$ | 800 | 5.8 | 85 | 54 | 31.3 |
| UF3 perm | 100 kDa | 1 | 230 | 27 | - | - | - |
|  | 30 kDa | 0.25 | 640 | 39 | - | - | - |
|  | 10 kDa | 0.31 | 1180 | 17.5 | 435 | 46 | 19.7 |
|  | 10 kDa | $1^1$ | 1180 | 5.4 | 435 | 14.3 | 6.1 |
| DMF2 effl. | 10 kDa | $1^1$ | 1950 | 3.2 | - | 12.4 | 5.3 |

*lmh = L/m²-h; $^1$Assumed value

As can be observed there is a significant difference between the "$t_r$" values for MFI-UF values without and with correction for flux. Temperature affects the estimated time. The lower the water temperature, the shorter the estimated time ($t_r$) will be.

## 8.5.2.2    Site B

### 8.5.2.2.1    Flux effect

The MFI values of RO feed water were determined at various flux rates. Results are illustrated in Figure 8.12.

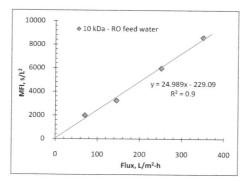

**Figure 8.12. MFI values vs. Flux for RO feed with a 10 kDa membrane - Site B**

A linear relationship was found between MFI and applied flux for RO feed water. The projected MFI value at 15 L/m²-h is ~430 s/L².

### 8.5.2.2.2    Particles/Colloids deposition on RO membrane

The measure particles and colloids deposition factors in the RO unit are presented in Table 8.14. The deposition factors were measured with 100, 50, 10 and 5 kDa membranes.

**Table 8.14. Measured particles/colloids deposition factor ($\Omega$) in RO unit**

| Sampling point | 100 kDa | 50 kDa | 10 kDa | 5 kDa* |
|---|---|---|---|---|
| RO feed | 980 | 2350 | 5975 | 2900 |
| RO conc | 1650 | 3850 | 10170 | 4300 |
| Deposition factor, $\Omega$ | -0.03 | 0.04 | -0.05 | 0.20 |
| $\Omega$ with 10 % IScf | 0.23 | 0.29 | 0.20 | 0.45 |

*Sample measured at 15 L/m²-h. The other values were measured at 250 L/m²-h.

For 100, 50 and 10 kDa membranes tested, the measured deposition factors are close to 0. This indicates that particles bigger that 10 kDa are not accumulating on the surface of the RO membranes. For 5 kDa membrane the deposition factor is about 20 %.

### 8.5.2.2.3    Cleaning frequency

The projected values for cleaning frequency for RO feed water and for raw water are presented in Table 8.15. For these projections, the conditions described in Table 8.11 were considered. Nevertheless, in the case where the measured deposition factors were not significant, a worse scenario was considered (full deposition of particles, $\Omega=1$).

For RO feed water, in the worse case it would take 3.3 months to reach a 15 % increase in NDP.

**Table 8.15. Predicted cleaning frequency in RO unit for particulate fouling - Site B**

| Sample | MWCO, kDa | $\Omega$ | MFI @ 250 lmh, s/L$^2$ | $t_{r\ @\ 250\ lmh}$, months | MFI @ 15 lmh, s/L$^2$ | $t_{r\ @\ 15\ lmh}$, months | 1 bar, $t_{r\ @\ 15\ lmh}$, months |
|---|---|---|---|---|---|---|---|
| RO feed | 100 | 0.23 | 980 | 42.7 | 105 | 398 | 94 |
|  | 50 | 0.29 | 2,350 | 14.1 | 290 | 114 | 27 |
|  | 10 | 0.20 | 5,975 | 8.1 | 430 | 112 | 26.5 |
|  | 5 | 0.45 | - | - | 2,900 | 7.4 | 1.7 |
| RO feed[1] | 100 | 1 | 980 | 9.8 | 105 | 91.6 | 21.7 |
|  | 50 | 1 | 2,350 | 4.1 | 290 | 33.2 | 7.8 |
|  | 10 | 1 | 5,975 | 1.6 | 430 | 22.4 | 5.3 |
|  | 5 | 1 | - | - | 2,900 | 3.3 | 0.8 |
| Raw water[2] | 100 | 1 | 25,340 | 0.38 | 2,816 | 3.4 | 0.8 |
|  | 50 | 1 | 44,285 | 0.22 | 4,921 | 2.0 | 0.5 |
|  | 10 | 1 | 64,500 | 0.15 | 7,167 | 1.3 | 0.3 |

[1]Worst case scenario ($\Omega=1$). [2]In case raw water is fed directly to RO units and considering $\Omega=1$.

In the case that raw water would be fed directly into the RO, the predicted times for 15 % increase in NDP are ~1 month.

### 8.5.2.3    Site C

#### 8.5.2.3.1    Flux effect

The MFI values of raw water and pre-treated water were measured at various flux rates. The raw water was tested with 100 and 10 kDa membranes and UF permeate and Coag+DMF effluent were measured with a 10 kDa membrane. The results are plotted in Figure 8.13.

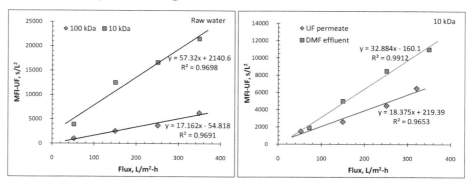

**Figure 8.13. MFI-UF vs. Flux for raw seawater (left) and pre-treated water (right)**

For the three water samples a linear trend was observed between flux and MFI. From the linear equations, the MFI-UF values at 15 L/m²-h (similar to SWRO operation) were projected. For raw water the projected value with the 10 kDa membrane (203 s/L²) is 15 times higher than that with the 100 kDa (3,000 s/L²).

The measured MFI-UF values for DMF effluent are higher than the MFI-UF values for UF permeate. The slopes of the equations are important to notice, as the rate of MFI-UF change with flux is much higher with DMF effluent than with UF permeate (~1.8 times). The projected MFI values for 15 L/m²-h are 495 s/L² for UF permeate and 333 s/L² for DMF effluent.

Even though the measured MFI UF values for DMF effluent were higher than the UF permeate values, the projected MFI-UF values at 15 L/m².h showed the opposite; DMF water's particulate fouling is lower than that of UF permeate.

### 8.5.2.3.2 Cleaning frequency

As the installation does not have a RO system, typical values from full scale operation were adopted and in some cases assumptions made considering worst scenarios. A recovery ($R$) of 40 % was considered (see Table 8.11).

The results of the model for UF permeate, Coag+DMF effluent and for Raw seawater are presented in Table 8.16. The projections were performed considering a RO unit with recovery = 45 % and flux = 15 L/m²-h. Also presented in the same table are the "$t_r$" results with and without correction for flux (250 and 15 L/m²-h).

**Table 8.16. Estimated times for 15 % increase in NDP for raw seawater, UF permeate and Coag+DMF effluent - Site C**

| Sample | MWCO | MFI @ 250 lmh, s/L² | $t_r$ @ 250 lmh, months | MFI @ 15 lmh, s/L² | $t_r$ @ 15 lmh, months | 1 bar, $t_r$ @ 15 lmh, months |
|---|---|---|---|---|---|---|
| Raw seawater | 100 kDa | 3,600 | 2.1 | 210 | 35.5 | 13.5 |
| | 30 kDa | 11,000 | 0.7 | - | - | |
| | 10 kDa | 13,000 | 0.6 | 3,000 | 2.5 | 0.9 |
| UF permeate | 100 kDa | 1,050 | 7.1 | - | - | |
| | 30 kDa | 2,900 | 2.6 | - | | |
| | 10 kDa | 8,200 | 0.9 | 500 | 14.9 | 5.7 |
| Coag+DMF effl. | 100 kDa | 300 | 24.8 | - | - | |
| | 30 kDa | 3,100 | 2.4 | - | | |
| | 10 kDa | 4,100 | 1.8 | 330 | 22.6 | 8.6 |

As can be observed there is a significant difference between the "$t_r$" values for MFI-UF values without and with correction for flux. The difference is between 5 and 15 times higher after flux correction. Furthermore, the 10 kDa membranes produce the shorter "$t_r$" times as the measured MFI values are higher than for 30 or 100 kDa.

Due to the higher MFI-UF values in the raw water the "$t_r$" values are estimated to be shorter than for the other two waters (5-9 times for 10 kDa membrane). UF permeate would foul the RO membranes in 15 months compared with 22.6 months for Coag+DMF effluent.

Temperature affects the estimated time. The lower the water temperature, the shorter the estimated time ($t_r$) will be.

In general terms, in practice the RO cleaning frequency is 2-3 times per year (4-6 months). It may also happen only once a year. Unfortunately, the pilot plant has no RO unit to compare with.

## 8.6    Conclusions

The fouling potential of raw seawater at three locations measured as MFI-UF with membranes with different pore sizes, showed large differences.

Fouling potential results in theoretical fouling rates, assuming deposition factor DF = 1, of 0.2 bar/month to ~1 bar/month depending on the pore size of the membranes used for MFI measurements.

The fouling potential of these raw waters are substantially reduced by conventional pre-treatment systems and ultrafiltration: for conventional pre-treatment 37 % - 74 % and ultrafiltration 60 % - 95 % depending on the location and MFI pore size.

Measured deposition factors varied between 0 and 1, depending on location and MFI pore size, which indicates differences in properties of the particles present.

Deposition factors below 1 result in substantially lower rates of fouling and cleaning frequencies. It is recommended to measure DF as many times as possible in operating plants.

## 8.7    References

BELFORT, G., DAVIS, R. H. & ZYDNEY, A. L. (1994). The behavior of suspensions and macromolecular solutions in crossflow microfiltration. *Journal of Membrane Science*, 96, 1-58.

BELFORT, G. & MARX, B. (1979). Artificial particulate fouling of hyperfiltration membranes II. Analyses and protection from fouling. *Desalination*, 28, 13-30.

BOERLAGE, S. F. E. (2001). *Scaling and Particulate Fouling in Membrane Filtration Systems*, Lisse, Swets&Zeitlinger Publishers.

BOERLAGE, S. F. E. (2007). Understanding the SDI and Modified Fouling Indices (MFI0.45 and MFI-UF). *IDA World Congress On Desalination and Water Reuse 2007 - Desalination: Quenching a Thirst* Maspalomas, Gran Canaria - Spain.

BOERLAGE, S. F. E., KENNEDY, M., TARAWNEH, Z., FABER, R. D. & SCHIPPERS, J. C. (2004). Development of the MFI-UF in constant flux filtration. *Desalination,* 161, 103-113.

BOERLAGE, S. F. E., KENNEDY, M. D., ANIYE, M. P. & SCHIPPERS, J. C. (2003). Applications of the MFI-UF to measure and predict particulate fouling in RO systems. *Journal of membrane science,* 220, 97-116.

BUSCH, M., CHU, R. & ROSENBERG, S. (2010). Novel Trends in Dual Membrane Systems for Seawater Desalination: Minimum Primary Pretreatment and Low Environmental Impact Treatment Schemes. *IDA Journal,* 2, 56-71.

CHONG, T. H., WONG, F. S. & FANE, A. G. (2007). Enhanced concentration polarization by unstirred fouling layers in reverse osmosis: Detection by sodium chloride tracer response technique. *Journal of Membrane Science,* 287, 198-210.

CHONG, T. H., WONG, F. S. & FANE, A. G. (2008). Implications of critical flux and cake enhanced osmotic pressure (CEOP) on colloidal fouling in reverse osmosis: Experimental observations. *Journal of Membrane Science,* 314, 101–111.

DOW (2005). *FILMTEC™ Reverse Osmosis Membranes - Technical Manual* DOW.

FRITZMANN, C., LÖWENBERG, J., WINTGENS, T. & MELIN, T. (2007). State-of-the-art of reverse osmosis desalination. *Desalination,* 216, 1-76.

HOEK, E. M. V. & ELIMELECH, M. (2003). Cake-Enhanced Concentration Polarization: A New Fouling Mechanism for Salt-Rejecting Membranes. *Environmental Science and Technology,* 37, 5581-5588.

HOEK, E. M. V., HONG, S. & ELIMELECH, M. (2001). Influence of membrane surface properties on initial rate of colloidal fouling of reverse osmosis and nanofiltration membranes. *Journal of Membrane Science,* 188, 115-128.

KHEDR, M. G. (2000). Membrane fouling problems in reverse osmosis desalination applications. *Desalination & Water Reuse,* 10, 8-17.

KOVALSKY, P., BUSHELL, G. & WAITE, T. D. (2009). Prediction of transmembrane pressure build-up in constant flux microfiltration of compressible materials in the absence and presence of shear. *Journal of Membrane Science,* 344, 204-210.

KOVALSKY, P., WANG, X., BUSHELL, G. & WAITE, T. D. (2008). Application of local material properties to prediction of constant flux filtration behaviour of compressible matter. *Journal of Membrane Science,* 318, 191-200.

RIETEMA, K. (1953). Stabilizing effects in compressible filter cakes. *Chemical Engineering Science,* 2, 88-94.

RUTH, B. F. (1935). Studies in Filtration III. Derivation of General Filtration Equations. *Industrial & Engineering Chemistry,* 27, 708-723.

SCHIPPERS, J. C. (1989). *Vervuiling van hyperfiltratiemembranen en verstopping van infiltratieputten,* Rijswijk, Keuringinstituut voor waterleidingartikelen KIWA N.V.

SCHIPPERS, J. C., HANEMAAYER, J. H., SMOLDERS, C. A. & KOSTENSE, A. (1981). Predicting flux decline or reverse osmosis membranes. *Desalination,* 38, 339-348.

SNF. 2009. *Water soluble polymers* [Online]. http://www.snfinc.com/cpsolutionscationic.html. [Accessed 14 December 2009].

TIEN, C. & BAI, R. (2003). An assessment ofthe conventional cake filtration theory. *Chemical Engineering Science,* 58, 1323 - 1336.

TILLER, F. M. (1958). The Role of Porosity in Filtration Part 3: Variable-pressure-Variable-rate Filtration. *A.I.Ch.E. Journal,* 4, 170-174.

TILLER, F. M. & KWON, J. H. (1998). Role of Porosity in Filtration: Behavior of Highly Compactible XIII. Cakes. *A.I.Ch.E. Journal,* 44, 2159-2167.

WOLF, P. H., SIVERNS, S. & MONTI, S. (2005). UF membranes for RO desalination pretreatment. *Desalination,* 182, 293-300.

YIANTSIOS, S. G., SIOUTOPOULOS, D. & KARABELAS, A. J. (2005). Colloidal fouling of RO membranes: an overview of key issues and efforts to develop improved prediction techniques. *Desalination,* 183, 257-272.

YIM, S. S., SONG, Y. M. & KWON, Y.-D. (2003). The Role of Pi, Po, and Pf in Constitutive Equations and New Boundary Conditions in Cake Filtration. *Korean J. of Chem. Eng.,* 20, 334-342.

# Chapter 9

## 9   Conclusions

## 9.1   Conclusions

Particulate/colloidal and organic fouling in seawater reverse osmosis (SWRO) systems results in flux decline, higher energy costs, increased salt passage, increased cleaning frequency, and use of chemicals. In practice, indices like SDI and MFI are used to assess particulate fouling, but they are performed at very high initial flux ($> 1500$ L/m$^2$-h) and do not take into account the deposition of particles/colloids in RO systems.

In this study, the Modified Fouling Index with ultrafiltration membranes (MFI-UF) at constant flux was further developed by incorporating the effects of particle/colloidal deposition and flux effect on particle rearrangement.

A new semi-portable set-up has been successfully developed to perform MFI-UF tests at constant flux filtration. The set-up has been used for on-site testing and for testing in laboratory. MFI-UF constant flux has potentially applications in: predicting the rate of fouling on a RO/NF membrane surface due to deposition of particles; verifying performance of MF/UF systems on the removal of colloidal matter; predicting rate of pressure increase in MF/UF systems within a filtration cycle; and verifying membrane integrity of MF/UF/NF/RO membrane systems.

Three important factors related with the use of the Modified Fouling Index ultrafiltration at constant flux as a tool to measure particulate fouling potential of a water and as a tool to estimate the rate of fouling in RO systems have been studied.

1.  The pore size or "molecular weight cut-off" (MWCO) of the membrane to be used in the test greatly influences the measured values. Furthermore, the MWCO of the membrane should be as close as possible to the pore size of RO membranes if the measured values will be used for fouling prediction.

2.  The formation of the fouling layer in the RO system or the deposition / accumulation of particles on the surface of the membranes. In the MFI model, this difference is considered by including the cake ratio factor in the prediction model and in practice is controlled by the flux rate at which filtration occurs.

3.  The filtration mode of the MFI test in comparison with the filtration mode of real RO systems (dead-end versus cross flow). This is site specific for each RO plant as it depends on the operational recovery, flux and the water characteristics (particle size distribution in the water). In the MFI prediction model, this is considered by measuring on-site the particle deposition factor in real RO plants.

## Flux effect (cake ratio factor)

A significant effect of filtration flux on the fouling potential was found. Consequences of this effect are the following: i) in reverse osmosis systems, the fouling potential at low flux drops dramatically; and ii) in ultrafiltration systems, the rate of fouling increases at high fluxes in particular when flux $>$ 60 $L/m^2$-h.

This effect was observed due to: i) the effect of compression in the cake layer occurring even at low flux rates (e.g., 20 $L/m^2$-h); ii) the effect of flux on rearrangement of particles during cake formation occurring above a certain flux value. In the case of the tested seawater, this value was around 60 $L/m^2$-h. At low flux rates, the effect of flux is not clear.

The observed effect of flux on the fouling potential has significant implications in fouling potential measurements like SDI, $MFI_{0.45}$ and MFI-UF constant flux. SDI and $MFI_{0.45}$ operate at constant pressure (2 bar) which yields high initial flux rates ($> 1500$ $L/m^2$-h). As a consequence overestimation of the fouling potential may occur.

The MFI-UF constant flux can operate at any flux rate (10-350 $L/m^2$-h). This is an advantage in considering the flux effect on fouling potential. To measure realistically the particulate fouling potential, the test should be performed at same flux as RO systems ($\sim$20 $L/m^2$-h) and MF/UF systems (60-80 $L/m^2$-h).

## Deposition of particles/colloids in RO systems

Without the deposition factor is not possible to estimate the rate of fouling in RO systems. The increase in ionic strength due to rejection of ions was found to influence the measured MFI values in RO concentrate water. In general, an increase in MFI-UF values with increasing salinity level in the solution was observed. For a feedwater with around 3.5% salinity and considering a RO recovery of around 40 %, the measured MFI-UF increase was $\sim$15 % with respect to the MFI value in the RO feed.

Correcting for ionic strength effect in the RO concentrate's MFI value increases the value of the deposition factor. Measured deposition factors varied between 0 and 1, depending on location and MFI pore size, which indicate differences in properties of the particles present.

## Fouling potential

A model equation to predict particulate fouling was further developed to incorporate the effects of particle/colloid deposition and flux. The fouling potential of raw seawater at three locations measured as MFI-UF with membranes with different pore sizes, showed large differences. Fouling potential results in theoretical fouling rates, assuming deposition factor $\Omega = 1$, of 0.2 bar/month to $\sim$1 bar/month depending on the pore size of the membranes used for MFI measurements.

The fouling potential of these raw waters are substantially reduced by conventional pre-treatment systems and ultrafiltration: for conventional pre-treatment 37-74 % and ultrafiltration 60-95 % depending on the location and MFI pore size.

Deposition factors below 1 result in substantially lower rates of fouling and cleaning frequencies. It is recommended to measure the deposition factor as many times as possible in operating plants.

**Organic matter characterization and organic foulants**

Seawater and estuarine water organic matter were analytically characterized by size exclusion chromatography couple with organic carbon detection and by fluorescence excitation-emission matrix. In the case of seawater (8 different locations), on average a DOC concentration of 1.08 mg-C/L was found, humic substances represented ~65 %, biopolymers ~12 %, and neutrals the remaining 23 %. In case of estuarine water (one site), on average a DOC concentration of 5.2 mg-C/L was found, humic substances consisted of ~72 %, biopolymers ~10 %, and neutrals the remaining 18 %.

In terms of pre-treatment assessment, beachwells and infiltration galleries (subsurface intakes) removed almost twice the biopolymer concentration (~70 %) in comparison with conventional (coagulation + media filtration) pre-treatment and membrane pre-treatment. Ultrafiltration units removed nearly 70 % of the biopolymers that were fed to the RO membranes.

The deposition factors and deposition rates of organic matter revealed that mainly biopolymers would deposit/accumulate on the surface of the RO membranes, which confirms organic matter is an important foulant in RO desalination systems.

In conclusion, the Modified Fouling Index with ultrafiltration membranes (MFI-UF) at constant flux was further developed by incorporating the effects of particle/colloidal deposition and flux. A new portable set-up was developed capable of working with membranes of various pore sizes (10-100 kDa) and flux ranges between 10-350 L/m$^2$-h. A model equation to predict particulate fouling was further developed to incorporate the effects of particle/colloid deposition and flux. Employing the new improved model, the rate of particulate/colloidal fouling potential of pre-treated seawater was found to be close to that of full scale desalination plants (between 0.2-1 bar/month), using a 10 kDa membrane at similar flux rate to a real RO system. The new developments presented in this study will enable engineers, plant operators and scientists not only to design better plants, but also to improve operation and monitoring of organic and particulate/colloidal fouling in SWRO systems.

# Summary

## PARTICULATE AND ORGANIC MATTER FOULING OF SWRO SYSTEMS: CHARACTERIZATION, MODELLING AND APPLICATIONS

### S.G. SALINAS RODRÍGUEZ

Fouling is a major concern in micro/ultra/nanofiltration and reverse osmosis systems in drinking water production from freshwater or from seawater. It is acknowledged in practice that the control of organic matter and particulate fouling is fundamental in decreasing costs related to membrane filtration independently of the applications. The control of organic matter and particulate fouling in membrane filtration systems can be improved by a clearer understanding of the processes involved in these phenomena, and by more accurate methods to predict and prevent these phenomena.

The objectives of this study are: to *i)* characterize bulk organic matter in seawater and bay water by various analytical techniques and link these measurements with fouling in membrane systems, and *ii)* further develop the Modified Fouling Index with ultrafiltration membranes at constant flux filtration as an accurate predictive tool to determine the particulate fouling potential of a feed water.

The final goal is to achieve a better knowledge of organic matter fouling and particulate matter fouling, that should enable engineers, plant operators and scientists not only to design better plants, but also to develop more effective tools for plant operation and monitoring of fouling.

This dissertation is organized in nine chapters. The first chapter corresponds to the introduction of the study. The last chapter summarizes the major conclusions of the study.

Chapters 2 and 3 are dedicated to organic matter characterization and applications in seawater full scale plants. Chapter 2 deals with the testing protocols and applications for mapping of organic matter components through liquid chromatography and fluorescence spectroscopy under high ionic strength conditions including parallel factor analysis and principal components analysis for seawater and estuarine water samples.

Chapter 3 makes use of the laboratory techniques described in chapter 2 to identify organic foulants in seawater, estuarine and bay sources for reverse osmosis plants. Several locations in Europe were studied.

Particulate/colloidal fouling potential is studied in the other five chapters (4, 5, 6, 7 and 8). Chapter 4 is the introduction to particulate/ colloidal fouling indices and presents a review of the current status of fouling indices used in seawater RO systems. Indices such as: Silt Density Index (SDI), Modified Fouling Index (MFI), MFI-UF constant pressure, MFI-UF constant flux and cross flow sampler (CFS) coupled with MFI-UF are discussed.

Chapter 5 presents the set-up and method development and applications related to the modified fouling index with ultrafiltration membranes at constant flux filtration (MFI-UF). The chapter characterizes the proposed membranes, describes the testing procedure for MFI-UF constant flux measurements and defines the limit of detection of the test. Applications related to comparison of various raw waters, particle size distribution, plant profiling, pre-treatment assessment and RO particulate fouling prediction are presented.

Chapter 6 studies the effect of flux rate on cake compression and on arrangement of particles in membrane filtration and on fouling indices.

Chapter 7 studies the particle deposition/accumulation in seawater reverse osmosis systems by measuring the particle deposition factor based on the MFI-UF constant flux measurements. A correction factor is proposed to consider effect of ionic strength on MFI values of RO concentrate.

Chapter 8 presents applications of the MFI-UF constant flux in pre-treatment assessment and in particulate fouling prediction.

Seawater and estuarine water organic matter were analytically characterized by size exclusion chromatography coupled with organic carbon detection and by fluorescence excitation-emission matrix. In the case of seawater (8 different locations), on average, a DOC concentration of 1.1 mg-C/L was found, humic substances represented ~65 %, biopolymers ~12 %, and neutrals the remaining 23 %. In the case of estuarine water (one site), on average a DOC concentration of 5.2 mg-C/L was found, humic substances consisted of ~72 %, biopolymers ~10 %, and neutrals the remaining 18 %.

In terms of pre-treatment assessment, beachwells and infiltration galleries (subsurface intakes) removed almost twice the biopolymer concentration (~70 %) in comparison with conventional (coagulation + media filtration) pre-treatment and membrane pre-treatment. Ultrafiltration units removed nearly 70 % of the biopolymers that were fed to the RO membranes. The deposition factors and deposition rates of organic matter revealed that mainly biopolymers would deposit/accumulate on the surface of the RO membranes, which suggests that organic matter fouling is important in RO desalination plants.

A new semi-portable set-up was successfully developed to perform MFI-UF tests at constant flux filtration. The set-up has been used for on-site testing and for testing in the laboratory. MFI-UF constant flux has potential applications in: predicting the rate of fouling on a RO/NF membrane due to deposition of particles; verifying performance of MF/UF systems on the removal of colloidal matter; predicting rate of pressure increase in MF/UF systems within a filtration cycle; and verifying membrane integrity of MF/UF/NF/RO membrane systems.

Three important factors related to MFI-UF constant flux as a tool to measure particulate fouling potential of a water and to estimate the rate of fouling in RO systems have been studied.

1. The pore size or "molecular weight cut-off" (MWCO) of the membrane to be used in the test influences greatly the measured values. Furthermore, the MWCO of the membrane should be as close as possible to the pore size of RO membranes if the measured values will be used for fouling prediction.

2. The formation of the fouling layer in the RO system or the deposition / accumulation of particles on the surface of the membranes. In the MFI model, this difference is considered by including the cake ratio factor in the prediction model and in practice is controlled by the flux rate at which filtration occurs.

3. The filtration mode of the MFI test in comparison with the filtration mode of real RO systems (dead-end versus cross flow). This is site specific for each RO plant as it depends on the operational recovery, flux and the water characteristics (particle size distribution in the water). In the MFI prediction model, this is considered by measuring on-site the particle deposition factor in real RO plants.

A significant effect of the filtration flux on the fouling potential was found. Consequences of this effect are the following: i) in reverse osmosis systems, the fouling potential at low flux drops dramatically; and ii) in ultrafiltration systems, the rate of fouling increases at high fluxes in particular when flux $>$ 60 L/m$^2$-h. This effect was observed due to: i) the effect of compression in the cake layer occurring even at low flux rates (e.g., 20 L/m$^2$-h); ii) the effect of flux on rearrangement of particles during cake formation occurring above a certain value. In case of the tested seawater, this value was around 60 L/m$^2$-h. At low flux rates, the effect of flux is not clear.

The observed effect of flux on the fouling potential has significant implications measurements like SDI, MFI$_{0.45}$ and MFI-UF constant flux. SDI and MFI$_{0.45}$ operate at constant pressure (2 bar) which yields high initial flux rates ($>$ 1500 L/m$^2$-h). As a consequence overestimation of the fouling potential may occur. The MFI-UF constant flux can operate at any flux rate (10-350 L/m$^2$-h). To measure realistically the particulate fouling potential, the test should be performed at the same flux as RO systems (~20 L/m$^2$-h) and MF/UF systems (60-80 L/m$^2$-h).

The increase in ionic strength due to rejection of ions was found to influence the measured MFI values in RO concentrate water. In general, an increase in

MFI-UF values with increasing the salinity level in the solution was observed. For a feedwater around 3.5 ‰ salinity and considering a RO recovery of around 40 %, the measured MFI-UF increase was ~15 % with respect to the MFI value in the RO feed. Correcting for ionic strength effect in the RO concentrate's MFI value increases the value of the deposition factor. Measured deposition factors varied between 0 and 1, depending on location and MFI pore size, which indicate differences in properties of the particles present.

The fouling potential of raw seawater at three locations measured as MFI-UF with membranes with different pore sizes, showed large differences. Fouling potential results in theoretical fouling rates, assuming deposition factor $\Omega = 1$, of 0.2-1 bar/month depending on the pore size of the membranes used for MFI measurements. The fouling potential of these raw waters are substantially reduced by conventional pre-treatment systems and ultrafiltration: for conventional pre-treatment 37 % - 74 % and ultrafiltration 60 % - 95 % depending on the location and MFI pore size. Deposition factors below 1 result in substantially lower rates of fouling and cleaning frequencies. It is recommended to measure DF as many times as possible in operating plants.

The new developments presented in this study will enable engineers, plant operators and scientists not only to design better plants, but also to improve operation and monitoring of organic and particulate/colloidal fouling in SWRO systems.

# Samenvatting

## MEMBRAANVERVUILING DOOR DEELTJES EN ORGANISCH MATERIAAL IN OMGEKEERDE OSMOSESYSTEMEN VOOR ZEEWATER: KARAKTERISATIE, MODELLERING EN TOEPASSINGEN

### S.G. SALINAS RODRÍGUEZ

Membraanvervuiling vormt een kritiek punt bij het gebruik van micro-/ultra-/nanomembranen en omgekeerde osmose (ook wel hyperfiltratie of reverse osmose genoemd) voor de productie van drinkwater uit zoet of zeewater. Het is algemeen bekend dat het beperken van organisch materiaal en deeltjesvervuiling essentieel is voor het verlagen van de kosten van membraanfiltratie, onafhankelijk van de toepassing. Het beperken van organisch materiaal en deeltjesvervuiling in membraanfiltratiesystemen kan worden verbeterd middels een beter begrip van de processen gerelateerd aan deze verschijnselen en middels accuratere methoden om deze verschijnselen te voorspellen en te voorkomen.

De specifieke doelstellingen van dit onderzoek zijn om: *i)* organisch materiaal in zeewater en baaiwater te karakteriseren middels verschillende analysetechnieken en vervolgens de meetresultaten te koppelen aan het optreden van membraanvervuiling, en *ii)* het verder ontwikkelen van de gemodificeerde vervuilingsindex (MFI) met ultrafiltratiemembranen (UF-membranen) bij constante flux tot een accurate test voor het voorspellen van het deeltjesvervuilingspotentieel van een voedingsstroom.

Het uiteindelijke doel is om een beter inzicht te krijgen in het mechanisme van organische vervuiling en vervuiling door deeltjes zodat het niet alleen mogelijk wordt voor ingenieurs, bedrijfstechnici en onderzoekers om betere installaties te ontwerpen, maar ook om betere methodes te ontwikkelen voor de besturing van de installaties en het volgen van de vervuilingsontwikkeling.

Dit proefschrift beslaat 9 hoofdstukken. Het eerste hoofdstuk gaat in op de introductie van het onderwerp. Het laatste hoofdstuk vat de belangrijkste conclusies van het onderzoek samen.

Hoofdstuk 2 and 3 zijn gewijd aan de karakterisatie van organisch materiaal en toepassingen in operationele zuiveringen voor zeewater. In hoofdstuk 2 wordt hierbij ingegaan op testprotocollen en het karakteriseren van organisch materiaal middels vloeistofchromatografie en fluorescerende spectroscopie bij hoge ionsterkte. Ook besproken worden de methoden van parallelle factor-analyse en hoofdcomponenten-analyse voor monsters die zowel uit zeewater als uit riviermondingen zijn genomen.

Hoofdstuk 3 past de labtechnieken uit hoofdstuk 2 toe om de organische vervuilende stoffen in watermonsters uit zee, riviermonding en zeearm te identificeren voor gebruik in omgekeerde osmose- (RO) installaties. Monsters zijn geanalyseerd vanuit meerdere locaties in Europa.

In de overige 5 hoofdstukken (4, 5, 6, 7 en 8) wordt ingegaan op het vervuilingspotentieel door deeltjes. Hoofdstuk 4 vormt hierbij een inleiding in indices voor deeltjesvervuiling, waarin een overzicht wordt gepresenteerd van de huidige vervuilingsindices zoals die worden toegepast bij RO-systemen. In dit hoofdstuk worden verschillende indices behandeld, waaronder de Silt Density Index (SDI), Modified Fouling Index (MFI), MFI-UF constante druk, MFI-UF constante flux en cross flow sampler (CFS) gekoppeld met MFI-UF.

Hoofdstuk 5 behandelt de opstelling en ontwikkeling van de methode en mogelijke toepassingen met betrekking tot de gemodificeerde vervuilingsindex met ultrafiltratiemembranen bij een constante flux-filtratie (MFI-UF). In dit hoofdstuk worden membranen gekarakteriseerd, de testprocedure voor MFI-UF constante flux-metingen beschreven en de detectielimieten besproken. Toepassingen worden besproken om verschillende onbehandelde waterstromen te kunnen vergelijken, alsmede de deeltjesgrootteverdeling, de profiling van de zuiveringen, het beoordelen van voorbehandelingen en het voorspellen van deeltjesvervuiling van RO membranen.

Hoofdstuk 6 gaat in op het effect van de flux op koekcompressie, op de rangschikking van deeltjes in membraanfiltratie en op vervuilingsindices.

Hoofdstuk 7 behandelt de deeltjesdepositie /-accumulatie in omgekeerde osmosesystemen voor zeewater door middel van het bepalen van de deeltjesdepositiefactor op basis van de MFI-UF constante flux-metingen. Een correctiefactor wordt geïntroduceerd om het effect van ionsterkte op MFI waarden voor het RO concentraat mee te nemen in de meting.

In hoofdstuk 8 wordt ingegaan op de mogelijke toepassingen van de MFI-UF constante flux op het beoordelen van voorbehandeling en op het voorspellen van mogelijke deeltjesvervuiling van membranen.

Zeewater en watermonsters uit riviermondingen zijn analytisch gekarakteriseerd middels chromatografie gebaseerd op deeltjesgrootte met organische koolstofdetectie en middels excitatie-emissie fluorescentiemicroscopie. Zeewater (van 8 verschillende locaties) bevat per liter gemiddeld 1.1 mg C, waarvan ~65 % bestaat uit humusachtige substanties, ~12 % uit biopolymeren, en ~23 % uit neutrale deeltjes. Water in riviermondingen (één locatie) bevat gemiddeld 5.2 mg C/L met ~72 % humus, ~10 % biopolymeren en ~18 % neutrale deeltjes.

Met betrekking tot het beoordelen van de voorbehandelingsmethoden kan gezegd worden dat strandputten en de infiltratiegalerie (onderoppervlakte-inname) samen zorgen voor een verwijdering van biopolymeren die bijna twee maal zo hoog is (~70 %) dan bereikt kan worden met conventionele (coagulatie + mediafiltratie) voorzuivering en membraanvoorzuivering.

Ultrafiltratie-eenheden verwijderden bijna 70 % van de biopolymeren die naar de RO membranen werden gestuurd. De depositiefactoren en -snelheden lieten zien dat het voornamelijk biolopolymeren zijn die neerslaan of ophopen op het membraanoppervlak; dit duidt op organische vervuiling.

Een nieuwe semi-draagbare installatie is met succes ontwikkeld om MFI-UF testen bij constante flux uit te kunnen voeren. De installatie is zowel in het laboratorium als op locatie getest. MFI-UF constante flux heeft potentiële toepassen in: het voorspellen van de vervuilingssnelheid van RO/UF systemen als gevolg van deeltjesneerslag; het verifiëren van de effectiviteit van MF/UF systemen in het verwijderen van colloïdaal materiaal; het voorspellen van de snelheid waarmee de druk toe zal nemen binnen een filtratiecyclus; en het vaststellen van de membraan-integriteit van MF/UF/NF/RO membraansystemen.

Drie belangrijke factoren zijn bestudeerd die gerelateerd zijn aan de MFI-UF constante flux als methode om het vervuilingspotentieel van een water te bepalen en om de vervuilingssnelheid in RO systemen te bepalen:

1. De poriegrootte of *"Molecular Weight Cut-off"* (MWCO) van het membraan dat gebruikt wordt in de test is van grote invloed op de gemeten waarde. Daarnaast moet de MWCO zo dicht mogelijk in de buurt liggen van de poriegrootte van het membraan indien de gemeten waarden gebruikt gaan worden voor een bepaling van het vervuilingspotentieel.

2. De vorming van de vervuilingslaag in RO systemen of de depositie / accumulatie van deeltjes op het membraanoppervlak. In het MFI model wordt hiermee rekening gehouden door het opnemen van een koekratiofactor in het voorspellingsmodel; in de praktijk wordt dit bepaald door de flux waarbij de filtratie plaatsvindt.

3. De filtratiemethode van de MFI-test in verhouding tot de filtratiemethode van het bestaande RO systeem (dead-end versus cross flow). Dit is locatie-specifiek omdat het afhankelijk is van de operationele terugwinningsfactor, de flux en de eigenschappen van het water (deeltjesgrootteverdeling in het water). In het MFI voorspellingsmodel wordt dit opgelost met het meten van de deeltjesdepositiefactor op locatie in de bestaande zuivering.

Een significant effect van de filtratieflux op het vervuilingspotentieel kon worden vastgesteld. De gevolgen hiervan zijn als volgt: i) in omkeerosmose-systemen zakt het vervuilingspotentieel aanzienlijk bij lage flux; ii) in ultrafiltratiesystemen neemt het vervuilingspotentieel toe bij een flux $> 60$

$L/m^2$-h. Dit effect werd waargenomen als gevolg van: $i$) het effect van koekcompressie dat ook bij lage flux optrad (e.g., 20 $L/m^2$-h); $ii$) het effect van de flux op het rangschikken van de deeltjes gedurende de vorming van de koek boven een bepaalde waarde. In het geval van het geteste zeewater lag deze waarde rond 60 $L/m^2$-h. Bij lagere fluxwaarden was het effect van de flux niet duidelijk.

Het effect van flux op het vervuilingspotentieel heeft significante gevolgen voor vervuilingspotentieelmetingen als SDI, $MFI_{0.45}$ and MFI-UF constante flux. SDI en $MFI_{0.45}$ worden bepaald bij constante druk (2 bar) waarmee hoge initiële fluxen bereikt worden ($>$ 1500 $L/m^2$-h). Hierdoor kan het vervuilingspotentieel mogelijk overschat worden. De MFI-UF constante flux kan bij elke flux uitgevoerd worden (10 - 350 $L/m^2$-h). Dit is een voordeel indien men bedenkt dat de flux een effect heeft op het vervuilingspotentieel. Om het deeltjesvervuilingspotentieel realistisch te kunnen meten dient de test uitgevoerd te worden bij dezelfde flux als de RO (~20 $L/m^2$-h) en MF/UF (60-80 $L/m^2$-h) systemen.

De toename in ionsterkte als gevolg van de membraanfiltratie bleek een effect te hebben op de gemeten MFI waarden in het RO concentraat: over het algemeen werd een toename waargenomen bij een verhoogd zoutgehalte in de oplossing. Voor de MFI van een voedingswater met ongeveer 3.5 ‰ en een RO terugwinningsfactor van rond de 40 % werd een toename van ~ 15 % waargenomen. Corrigerend voor het effect van de ionsterkte ziet men een stijging in de depositiefactor met de MFI van het RO concentraat. Gemeten depositiefactoren varieerden tussen 0 en 1, afhankelijke van de locatie en de MFI poriegrootte, wat duidt op verschillen in de eigenschappen van de aanwezige deeltjes.

Het vervuilingspotentieel van onbehandeld zeewater van drie locaties, gemeten als MFI-UF met membranen met verschillende poriegroottes, vertoonde grote verschillen. Bij een aangenomen depositiefactor van $\Omega = 1$ resulteerde dit in in theoretische vervuilingssnelheden voor het vervuilingspotentieel van 0.2-1 bar/maand, afhankelijk van de poriegrootte van de membranen die gebruikt werden voor de MFI bepalingen. Het vervuilingspotentieel van deze onbehandelde waterstromen nam aanzienlijk af met conventionele voorbehandelingstechnieken en ultrafiltratie: voor conventionele voorbehandeling 37 % - 74 % en voor ultrafiltratie 60 % - 95 % afhankelijk van de locatie en de MFI poriegrootte. Depositiefactoren lager dan 1 resulteren in substantieel lagere vervuilingssnelheden en bijbehorende schoonmaakfrequenties. Het is aan te bevelen de depositiefactor zo vaak mogelijk te bepalen in drinkwaterzuiveringen.

# Publications

## Journals

SALINAS RODRÍGUEZ, S. G., KENNEDY, M. D., AMY, G. & SCHIPPERS, J. C. (2011). The role of particles/colloids in SWRO systems - Fouling prediction and validation. *Water Research*, submitted.

SALINAS RODRÍGUEZ, S. G., KENNEDY, M. D., AMY, G. & SCHIPPERS, J. C. (2011). Particle deposition in SWRO systems. *Water Research*, submitted.

SALINAS RODRÍGUEZ, S. G., KENNEDY, M. D., AMY, G. & SCHIPPERS, J. C. (2011). The modified fouling index - ultra filtration - constant flux for seawater applications. *Water Research*, submitted.

SALINAS RODRÍGUEZ, S. G., KENNEDY, M. D., AMY, G. & SCHIPPERS, J. C. (2011). Flux effects on cake compression in membrane filtration. *Water Research*, submitted.

SALINAS RODRÍGUEZ, S. G., KENNEDY, M. D., AMY, G. & SCHIPPERS, J. C. (2011). A review of fouling indices used in RO systems. *Water Research*, submitted.

SALINAS RODRÍGUEZ, S. G., KENNEDY, M. D., AMY, G. L. & SCHIPPERS, J. C. (2011). Flux dependency of particulate/colloidal fouling in seawater reverse osmosis systems. *Desalination and Water Treatment*, in press.

SALINAS RODRÍGUEZ, S. G., EKOWATI, Y., KENNEDY, M. D., AMY, G. & SCHIPPERS, J. C. (2011). Fouling potential of organic foulants in reverse osmosis systems using Modified Fouling Index - Ultrafiltration constant flux. *Desalination*, to be submitted.

SALINAS RODRÍGUEZ, S. G., KENNEDY, M. D., SCHIPPERS, J. C. & AMY, G. L. (2009). Organic foulants in estuarine and bay sources for seawater reverse osmosis – Comparing pre-treatment processes with respect to foulant reductions. *Desalination and Water Treatment*, 9, 155-164.

SALINAS RODRÍGUEZ, S. G., AL-RABAANI, B., KENNEDY, M. D., SCHIPPERS, J. C. & AMY, G. L. (2009). MFI-UF constant pressure at high ionic strength conditions. *Desalination and Water Treatment*, 10, 64-72.

BAGHOTH, S. A., MAENG, S. K., SALINAS RODRIGUEZ, S. G., RONTELTAP, M., SHARMA, S., KENNEDY, M. D. & AMY, G. (2008). An Urban Water Cycle Perspective of Natural Organic Matter (NOM): NOM in Drinking Water, Wastewater Effluent, Storm water, and Seawater. *Water Science & Technology: Water Supply*, 8, 701-707.

SALINAS RODRÍGUEZ, S. G., KENNEDY, M. D., DIEPEVEEN, A., PRUMMEL, H. & SCHIPPERS, J. C. (2008). Optimization of PACl dose to reduce RO cleaning in an IMS. *Desalination*, 220, 239-251.

SALINAS RODRÍGUEZ, S. G., KENNEDY, M. D., PRUMMEL, H., DIEPEVEEN,
    A. & SCHIPPERS, J. C. (2008). PACl: A simulation of the change in Al
    concentration and Al solubility in RO. *Desalination*, 220, 305-312.


**Conference publications**


GASIA BRUCH, E., BUSCH, M., SALINAS RODRÍGUEZ, S. G. & KENNEDY, M.
    D. (2011). Improvement of fouling indices measurements and modeling of
    their relevance. In: IDA (ed.) *Desalination World Congress - Perth
    Convention and Exhibition Centre*. Perth: IDA.

SALINAS RODRÍGUEZ, S. G., KENNEDY, M. D., AMY, G. & SCHIPPERS, J. C.
    (2010). Flux dependency of particulate fouling by MFI-UF measurements in
    seawater reverse osmosis systems. In: EDS (ed.) *EuroMed 2010: Desalination
    for Clean Water and Energy*. Tel-Aviv, Israel: EDS.

SALINAS RODRÍGUEZ, S. G., KENNEDY, M. D., AMY, G. & SCHIPPERS, J. C.
    (2010). Flux dependency of particulate fouling in seawater reverse osmosis
    systems. In: EDS (ed.) *Membranes in drinking water production and waste
    water treatment*. Trondheim, Norway: EDS/IWA.

LOZIER, J. C., BANKSTON, A., BEATY, J., GARCIA-ALEMAN, J., SCHARF, R.,
    AMY, G. & SALINAS RODRÍGUEZ, S. G. (2009). Use of advanced NOM
    characterization methods to trace the fate of organic contaminants from a
    membrane backwash recycle scheme. *In:* AWWA, ed. *Membrane Technology
    Conference*, Memphis, Tennessee United States. AWWA and AMTA.

SALINAS RODRÍGUEZ, S. G., KENNEDY, M. D., SCHIPPERS, J. C. & AMY, G.
    L. (2009). Tracking organic matter in four SWRO plants – Comparing pre-
    treatment processes with respect to organic foulant reductions. *In:* DUT (ed.)
    *High quality drinking water conference*. Delft.

SALINAS RODRÍGUEZ, S. G., MAMOUN, A., SCHURER, R., KENNEDY, M. D.,
    AMY, G. L. & SCHIPPERS, J. C. (2009). Modified fouling index (MFI-UF)
    at constant flux for seawater RO applications. In: EDS (ed.) *Desalination for
    the Environment: Clean water and Energy*. Baden-Baden, Germany: European
    desalination society.

SALINAS RODRÍGUEZ, S. G., KENNEDY, M. D., SCHIPPERS, J. C. & AMY, G.
    (2008). Identification of organic foulants in estuarine and seawater reverse
    osmosis systems – Comparison for different pre-treatments. In: EDS-INSA
    (ed.) *Membranes in Drinking water production and waste water treatment*.
    Toulouse, France: European desalination society.

SALINAS RODRÍGUEZ, S. G., KENNEDY, M. D., CHENET, J. G., GARCIA
    ALEMAN, J., BANKSTON, A. & AMY, G. (2008). Tracking organic matter
    removal in a membrane treatment plant - Air scour vs. hydraulic backwash.
    *In:* IWA (ed.) *Natural organic matter: from source to tap*. Bath, UK: IWA -
    Cranfield University.

SALINAS RODRÍGUEZ, S. G., LI, S., CHENET, J., FUTSELAAR, H.,
    ABRAHAMSE, A., HEIJMAN, B., KENNEDY, M. D. & AMY, G. (2008).

Colloidal organic matter fouling of ultrafiltration membranes: Role of calcium & membrane properties. *In:* UNAM (ed.) *Young water professionals.* Mexico DF, Mexico: IWA-UNAM.

SALINAS RODRIGUEZ, S. G., GONZALES T, A., KENNEDY, M. & AMY, G. (2008). Fluorescence of selected organic matter compounds: looking at the effect of concentration, ionic strength and pH. *In:* BIRMINGHAM, U. O., ed. AGU Chapman. *Conference on Organic Matter Fluorescence,* Edgbaston, Birmingham, UK.

SALINAS RODRÍGUEZ, S. G., KENNEDY, M. D. & AMY, G. (2007). State of the art in NOM characterization techniques for water treatment applications. *In:* LEUVEN, U.-U. O., ed. *International Congress on Development, Environment and Natural Resources: Multi-level and Multi-scale Sustainability,* 11-14 July 2007 Cochabamba, Bolivia. UMSS-University of Leuven.

## Book chapters

KENNEDY, M. D., KAMANYI, J., SALINAS RODRÍGUEZ, S. G., LEE, N. H., SCHIPPERS, J. C. & AMY, G. (2008) Water treatment by microfiltration and ultrafiltration. IN LI, N. N., FANE, A. G., HO, W. & MATSUURA, T. (Eds.) *Advanced membrane technology and Applications.* New Jersey, John Wiley & Sons.

AMY, G. L., SALINAS RODRÍGUEZ, S. G., KENNEDY, M. D., SCHIPPERS, J. C., RAPENNE, S., REMIZE, P.-J., BARBE, C., MANES, C.-L. D. O., WEST, N. J., LEBARON, P., KOOIJ, D. V. D., VEENENDAAL, H., SCHAULE, G., PETROWSKI, K., HUBER, S., SIM, L. N., YE, Y., CHEN, V. & FANE, A. G. (2011). Water quality assessment tools. In: DRIOLI, E., CRISCUOLI, A. & MACEDONIO, F. (eds.) *Membrane-Based Desalination - An Integrated Approach (MEDINA).* IWA.

LATTEMANN, S., SALINAS RODRÍGUEZ, S. G., KENNEDY, M., SCHIPPERS, J. C. & AMY, G. (2011). Is seawater desalination green? - An evaluation of state of the art of pre-treatment and desalination technologies considering environmental and performance aspects. In: LIOR, N., BALABAN, M., DARWISH, M., MIYATAKE, O., WANG, S. & WILF, M. (eds.) *Advances in water desalination Vol. 1.* New Jersey: John Wiley & Sons.

SHARMA, S. K., BAGHOTH, S. A., MAENG, S. K., SALINAS RODRÍGUEZ, S. G. & AMY, G. L. (2011). Chapter 3. Natural organic matter: Characterization profiling as a basis for treatment process selection and performance monitoring. In: A. VAN NIEUWENHUIZEN & J. VAN DER GRAAF (eds.) *Handbook on particle separation processes.* IWA.

## Best oral presentation

International Congress on Development, Environment and Natural Resources: Multi-level and Multi-scale Sustainability. Cochabamba, Bolivia, San Simon Major

University - University of Leuven, (2007). *State of the art in NOM characterization techniques for water treatment applications*, S. Salinas Rodriguez, M.D. Kennedy, G.L. Amy.

# Curriculum Vitae

| | |
|---|---|
| *October 8ᵗʰ, 1978:* | Born in Oruro as SERGIO GENARO SALINAS RODRÍGUEZ. |
| *Feb. 1984 – Nov. 1995:* | Bachiller en humanidades, Colegio Don Bosco, Cochabamba - Bolivia. |
| *Feb. 1996 – Jul. 2001:* | Licentiate in Civil Engineering, Universidad Mayor de San Simon. Specialization in Structures and Sanitary engineering. Cochabamba - Bolivia. |
| *Jul. 2001 – Feb. 2002:* | Engineer. Various projects. Cochabamba - Bolivia. |
| *Feb. 2002 – Dec. 2002:* | Master in Irrigation and Drainage, Universidad Mayor de San Simon, Cochabamba - Bolivia. |
| *Jan. 2003 – Sep. 2004:* | Engineer. Sanitary and structures design in various projects. Civil works supervisor. Cochabamba - Bolivia. |
| *Oct. 2004 – Apr. 2006:* | Master of Science in Water Supply Engineering at UNESCO-IHE, The Netherlands. |
| *Apr. 2006 – Dec. 2006:* | Project assistant at Urban Water and Sanitation department, UNESCO-IHE, The Netherlands. |
| *Dec. 2006 – Dec. 2010:* | PhD student, Urban Water and Sanitation department, UNESCO-IHE, The Netherlands. |
| *Since Jan. 2011:* | Lecturer in Water Treatment Technology, Urban Water and Sanitation department, UNESCO-IHE, The Netherlands. |

T - #0143 - 160425 - C250 - 244/170/14 - PB - 9780415620925 - Gloss Lamination